Curriculum, Community, and Urban School Reform

SECONDARY EDUCATION IN A CHANGING WORLD

Series editors: Barry M. Franklin and Gary McCulloch

Published by Palgrave Macmillan:

The Comprehensive Public High School: Historical Perspectives
By Geoffrey Sherington and Craig Campbell
(2006)

Cyril Norwood and the Ideal of Secondary Education
By Gary McCulloch
(2007)

The Death of the Comprehensive High School?: Historical, Contemporary, and Comparative Perspectives
Edited by Barry M. Franklin and Gary McCulloch
(2007)

The Emergence of Holocaust Education in American Schools
By Thomas D. Fallace
(2008)

The Standardization of American Schooling: Linking Secondary and Higher Education, 1870–1910
By Marc A. VanOverbeke
(2008)

Education and Social Integration: Comprehensive Schooling in Europe
By Susanne Wiborg
(2009)

Reforming New Zealand Secondary Education: The Picot Report and the Road to Radical Reform
By Roger Openshaw
(2009)

Inciting Change in Secondary English Language Programs: The Case of Cherry High School
By Marilee Coles-Ritchie
(2009)

Curriculum, Community, and Urban School Reform
By Barry M. Franklin
(2010)

Curriculum, Community, and Urban School Reform

Barry M. Franklin

palgrave
macmillan

CURRICULUM, COMMUNITY, AND URBAN SCHOOL REFORM

First published in 2010 by
PALGRAVE MACMILLAN®
in the United States—a division of St. Martin's Press LLC,
175 Fifth Avenue, New York, NY 10010.

Where this book is distributed in the UK, Europe and the rest of the world, this is by Palgrave Macmillan, a division of Macmillan Publishers Limited, registered in England, company number 785998, of Houndmills, Basingstoke, Hampshire RG21 6XS.

Palgrave Macmillan is the global academic imprint of the above companies and has companies and representatives throughout the world.

Palgrave® and Macmillan® are registered trademarks in the United States, the United Kingdom, Europe and other countries.

ISBN: 978–0–230–61234–1

Library of Congress Cataloging-in-Publication Data is available from the Library of Congress.

A catalogue record of the book is available from the British Library.

Design by Newgen Imaging Systems (P) Ltd., Chennai, India.

First edition: March 2010

D 10 9 8 7 6 5 4 3 2

Printed in the United States of America.

For

Herb, Jośe, Tom, Bill, Mike, Gary, Miguel

Contents

Series Editor's Foreword ix

Acknowledgment xiii

List of Abbreviations xvii

1 Community and Curriculum: A Conceptual Framework for Interpreting Urban School Reform 1

2 Community Conflict and Compensatory Education in New York City: More Effective Schools and the Clinic for Learning 29

3 Community, Race, and Curriculum in Detroit: The Northern High School Walkout 57

4 Race and Community in a Black Led City: The Case of Detroit and the Mayoral Takeover of the Board of Education 81

5 Educational Partnerships, Urban School Reform, and the Building of Community 105

6 Educational Partnerships and Community: Education Action Zones and "Third Way" Educational Reform in Britain 143

7 Smaller Learning Communities and the Reorganization of the Comprehensive High School 175
Barry M. Franklin and Richard Nye

Epilogue Community in a Cosmopolitan World 209

Notes 215

Index 249

Series Editor's Foreword

Among the educational issues affecting policy makers, public officials, and citizens in modern, democratic, and industrial societies, none has been more contentious than the role of secondary schooling. In establishing the Secondary Education in a Changing World series with Palgrave Macmillan, our intent is to provide a venue for scholars in different national settings to explore critical and controversial issues surrounding secondary education. We envision our series as a place for the airing and hopefully resolution of these controversial issues.

More than a century has elapsed since Emile Durkheim argued the importance of studying secondary education as a unity, rather than in relation to the wide range of subjects and the division of pedagogical labor of which it was composed. Only thus, he insisted, would it be possible to have the ends and aims of secondary education constantly in view. The failure to do so accounted for a great deal of the difficulty with which secondary education was faced. First, it meant that secondary education was "intellectually disorientated," between "a past which is dying and a future which is still undecided," and as a result "lacks the vigor and vitality which it once possessed" (Durkheim 1938/1977, p. 8). Second, the institutions of secondary education were not understood adequately in relation to their past, which was "the soil which nourished them and gave them their present meaning, and apart from which they cannot be examined without a great deal of impoverishment and distortion" (10). And third, it was difficult for secondary school teachers, who were responsible for putting policy reforms into practice, to understand the nature of the problems and issues that prompted them.

In the early years of the twenty-first century, Durkheim's strictures still have resonance. The intellectual disorientation of secondary education is more evident than ever as it is caught up in successive waves of policy changes. The connections between the present and the past have become increasingly hard to trace and untangle. Moreover, the distance between policy makers on the one hand and the practitioners on the other has rarely seemed as immense as it is today. The key mission of the current series of books is, in the spirit of Durkheim, to address these underlying dilemmas of secondary education and to play a part in resolving them.

Curriculum, Community, and Urban School Reform by Barry M. Franklin, contributes to this mission by exploring urban school reform over the past half-century, particularly as it affects high schools, through the conceptual lens of community. Looking at high schools, and also including elementary and junior high schools, through the framework of community can enhance our understanding of how these institutions work and the challenges they face. The volume ranges widely across a number of detailed case studies to investigate how these ideas and practices play themselves out in different urban settings. The first is based in New York City and examines the efforts of the More Effective Schools Program and the Clinic for Learning to improve the academic achievement of disadvantaged minority youth. The scene then shifts to Detroit in Michigan, initially to study the case of the Northern High School walkout in 1966, and then to analyze the mayoral takeover of failing school districts announced in 1999. The next case study is based on three different venues, those of New York City, Detroit, and Minneapolis, to develop a comparative appraisal of educational partnerships and their role in school reform. The volume then develops this comparison further with a detailed examination of educational partnerships in the British context, specifically the case of the Education Action Zones introduced by Tony Blair's Labour government after 1997. Following this, it pursues the theme of community through an interpretation of smaller learning communities and the reorganization of the comprehensive high school. Finally, it broadens its lens to the fullest extent with reflections on community in a globalized context, that of a cosmopolitan world.

Nearly twenty-five years after the publication of Professor Franklin's first major study in the field, the influential work *Building the American Community: The School Curriculum and the Search for Social Control* the present volume therefore reflects a number of significant developments in outlook and approach. First of all, it attempts to take forward a social history analysis as opposed to a mainly intellectual history of community and curriculum. Second, it is more wide ranging and varied, both in the nature of the programs and specific cases in which it is interested and in the locales with which it is concerned. Indeed, while the central focus of the volume is on high schools, Franklin also makes reference to reforms that affect elementary and junior high schools. This broad ranging treatment also lends itself to a further characteristic, its international and global relevance quite apart from its significance within the United States, in which its analysis of cosmopolitanism will be a key topic of discussion and debate. Finally, Franklin's volume brings together the past and the present in a cohesive and integrated fashion to address the problems that face us today on the basis of our historical experience.

Curriculum, Community, and Urban School Reform is the ninth volume to be published in our series. It exemplifies well the combination of social, historical, and comparative approaches to secondary education that we have sought to emphasize throughout, and the potential scope of these for furthering our understanding of ideas, policies, and practices in education. As we see the trajectory of the series advancing during the next few years, our intent is to seek additional volumes that bring these issues still further to the attention of studies in secondary education.

GARY MCCULLOCH
Series Co-Editor

Reference

Durkheim, E. (1938/1977). *The Evolution of Educational Thought: Lectures on the Formation and Development of Secondary Education in France.* London: Routledge and Kegan Paul.

Acknowledgment

The idea of community has played an important role in my research and writing throughout my academic career. In this book, I explore this notion as a conceptual lens for understanding urban school reform since 1960 with particular reference to issues of curriculum. I should say at the outset that trying to craft a book about curriculum is something of a risky venture since this seemingly straightforward concept turns out to be contested terrain. For some, curriculum is defined simply as the subject matter that the schools teach. Others take a broader view and see the curriculum as encompassing a range of experiences from the planned to the unplanned that occur in schools. In the same vein, there are those who view the curriculum as explicit elements that affect what occurs in classrooms, while others include within the realm of curriculum certain implicit and hidden elements. In my work, I lean toward the broader view and see the curriculum as encompassing a variety of environmental factors that in different ways affect the content and processes of schooling. It is for this reason that I focus in some chapters on questions of school organization as well as the interplay between stakeholders within and outside the school. These are to my way of thinking important and necessary topics for exploring matters of curriculum.

There is a key difference between this book and my first effort in 1986 to examine the concept of community as an interpretive lens for understanding curriculum. That book, *Building the American Community: The School Curriculum and the Search for Social Control* (Falmer Press) was largely an intellectual history of the curriculum that examined, except for one chapter, proposals and recommendations for what the schools should teach. This present book does include discussions that are of the sort that might be thought of as constituting intellectual history. This is especially the case in chapter 1 where I make use of some postmodern categories to talk about community as representing a series of circulating discourses as well as my view that this book is more akin to a genealogy or a history of the present than a traditional history. For the most part, however, this book is a social history that is comprised of a number of distinct case studies or policy narratives of curriculum reform that have appeared on the scene during the last four decades of the twentieth century and now into the twenty-first.

In writing a book closer to the end of a career than to the beginning, one acquires a vast array of intellectual debts. For me, the seven individuals to whom I dedicate this book stand out as having provided me over the years with the thoughtful comments and criticisms that have in different and important ways helped to shape not only this book but my scholarship more broadly. They are Herbert Kliebard, Jośe Rosario, Thomas Popkewitz, Michael Apple, William Reese, Gary McCulloch, and Miguel Pereyra. Each in their own way, they have been teachers, colleagues, collaborators, and most importantly dear friends. Words cannot express my gratitude to them.

I wish to acknowledge the following institutions for granting me access to the manuscript collections used in this book and to their staffs for assisting me in my research: Bentley Historical Library, University of Michigan; Detroit Public School Archives, Detroit Board of Education; Robert F. Wagner Labor Archives, New York University; Special Collections, Milbank Memorial Library, Teachers College, Columbia University; San Francisco History Center, San Francisco Public Library; The Hoover Institution of War, Revolution and Peace, Stanford University; and Walter P. Reuther Library, Wayne State University.

The research that I undertook for this book was supported by a number of grants. I am indebted for this support to the Office of the Vice President for Research at Utah State University, the Horace Rackham Graduate School at the University of Michigan, the Office of Research at the University of Michigan-Flint, and the Spencer Foundation.

In conducting my research, I was aided by appointments during the Winter of 2001 as a Visiting Fellow in the Policy Studies Unit at the Institute of Education, University of London and during the Spring of 2001 as an Honorary Fellow in the Department of Curriculum and Instruction at the University of Wisconsin—Madison. I am especially grateful to faculty and staff at these two institutions for their support and help.

Throughout my work on this manuscript, my colleagues in the School of Teacher Education and Leadership (TEAL) in the Emma Eccles Jones College of Education and Human Services at Utah State University have been a constant source for ideas, insights, and suggestions that have made this a better volume. Three members of the faculty deserve special mention. TEAL's Associate Dean, Martha Dever, has over the years been supportive of my research and has provided me with the kind of teaching schedules and travel support that has made it possible for me to complete this volume. I have frequently relied on my colleague in the office across the hall from me at Utah State, Sherry Marx, for trying out this or that idea or concept that has informed this volume. Steven Camicia's research on cosmopolitanism has been especially helpful, as I have sought to situate the idea of community in the broader context of twenty-first century educational reform. Ultimately, of course, I assume total responsibility for this volume and the interpretation advanced within.

Two doctoral students in our program in curriculum and instruction in TEAL deserve mention. Richard Nye, my doctoral student and research assistant, played a major role as the co-author of our chapter on smaller learning communities (chapter 7). His coding of the interviews that comprised a major source of data for this chapter, his writing of sections of the chapter, and his knowledge of the research literature on smaller learning communities made invaluable contributions to the volume. Juan Juan Zhu took on the task of identifying and locating the various journals that were used in examining the history of small schools and classroom size in chapter 7.

I also wish to acknowledge invitations from the following institutions and organizations to present earlier versions of the chapters in this book as lectures and seminar papers: Institute of Education, University of London (UK); University of Granada (Spain); University of Zurich (Switzerland); Danish University School of Education (Denmark); Sociology of Education Section, Hungarian Sociological Association (Hungary); University of Pecs (Hungary); University of Birmingham (UK), University of East Anglia (UK); and Indiana University/Purdue University (United States).

Throughout my work on this book, I have benefited from the help and support provided to me by Julia Cohen, Associate Editor at Palgrave Macmillan and her editorial assistant, Samantha Hassey. They have done much to smooth the development of my ideas on community into a finished book.

No matter when and how it occurs, research and writing affect one's family and their lives together. As always, I am indebted to my wife Lynn Marie and my children Jeremy and Nathan for their willingness to give me the space to pursue this work. Lynn Marie deserves special mention. She is an excellent copy editor, and her reading of the book was critical in creating a coherent and well-written narrative.

Portions of this volume represent revisions that have appeared elsewhere. Chapter 3 originally appeared in *History of Education* 33 (March, 2004), 137–156 (*History of Education* is published by Taylor and Francis whose Web site is http://www.informaworld.com). Chapter 4 is a greatly revised, updated, and rewritten version of an earlier essay that was originally entitled "Race, Restructuring, and Educational Reform: The Mayoral Takeover of the Detroit Public Schools," which appeared in *Reinterpreting*

Urban School Reform: Have Urban Schools Failed, or Has the Reform Movement Failed Urban Schools?, edited by Louis F. Mirón and Edward P. St. John (Albany: State University of New York Press, 2003), pp. 95–125. Chapter 5 is a revised version of an essay that originally appeared under the title "Gone Before you Know It: Urban School Reform and the Short life of the Education Action Zone Initiative," which appeared in the *London Review of Education* 3 (March 2005), 3–27 (London Review of Education is published by Taylor and Francis whose Web site is http://www.informaworld.com). I appreciate the permission of these publishers to include this material.

Finally, I should say that my decision to place this book in the series on Secondary Education in a Changing World that I co-edit with Gary McCulloch was deliberate. Although I treat elementary school in some chapters, the book is largely a consideration of policies that in different ways affect secondary education and consequently its inclusion supports the purposes of this series.

Abbreviations

ACORN	Association of Community Organizations for Reform Now
ADNS	All Day Neighborhood Schools
AFL-CIO	American Federation of Labor and Congress of Industrial Organization
AYP	adequate yearly progress
BDP	Bernard Donovan Papers, Milbank Memorial Library, Teachers College, Columbia University
CAC	Citizens Advisory Committee on School Needs
CAC	Community Advisory Council
CCE	Center for Collaborative Education
CEI	Center for Educational Innovation
CEO	chief executive officer
CORE	New York City Congress of Racial Equality
DCCR	Detroit Commission on Community Relations Collection, Human Rights Department, Archives of Labor History and Urban Affairs, Walter P. Reuther Library, Wayne State University
DEA	Detroit Education Association
DfES	Department for Education and Skills
DFT	Detroit Federation of Teachers
DPSA	Detroit Public School Archives
DUL	Detroit Urban League Collection, Bentley Historical Library, University of Michigan
EAZ	Education Action Zones
EEO	Citizens Advisory Committee on Equal Educational Opportunity
EMO	educational management organization
ESL	English as a Second Language
GM	General Mills
ICT	information and communication technology
IT	information technology
JHS	Junior High School

LEA	Local Education Authorities
MES	More Effective Schools
MES	United Federation of Teachers Collection, More Effective Schools Program, Robert F. Wagner Labor Archives, New York University
NAACP	National Association for the Advancement of Colored People
NDP	Norman Drachler Papers, The Paul and Jean Hanna Archival Collection, The Hoover Institution of War, Revolution and Peace, Stanford University
NUT	National Union of Teachers
NYNSR	New York Networks for School Renewal
NYU	New York University
OFSTED	Office for Standards in Education
PEA	Public Education Association
RRP	Remus G. Robinson Papers, Archives of Labor History and Urban Affairs, Walter P. Reuther Library, Wayne State University
RS	Rose Shapiro Papers, Milbank Memorial Library, Teachers College, Columbia University
SREB	Southern Regional Education Board
STAR	State of Tennessee's Student/Teacher project
SURR	School Under Registration Review
UFT	United Federation of Teachers
UFT	United Federation of Teachers Collection, Robert F. Wagner Labor Archives, New York University
WEPIC	West Philadelphia Improvement Corps
WPA	Works Progress Administration

Chapter 1

Community and Curriculum: A Conceptual Framework for Interpreting Urban School Reform

The starting place for this volume is Timberton Central High School. A comprehensive high school of around 1,500 students with a growing Latino/a enrollment, the school is located in the economically distressed, high poverty, and racially diverse Intermountain West city that I am calling Timberton.[1] My research assistant and I spent several hours a week for about a year beginning in August of 2006 in this school as it underwent a reorganization into a number of smaller learning communities. Under this new arrangement the overall population of the school would remain about the same with each largely self-contained learning community enrolling from 200 to 400 students. The change would not alter the school's comprehensiveness but would replace the departmental organization that typifies such high schools with a Ninth Grade Center and four career oriented units—Applied Science and Technology, Arts and Humanities, Business and Computers, and Health Science and Human Service. Midway through our time in the school, it was decided to merge the Business and Computers Community with that of the Arts and Humanities Community to create three career oriented communities for the tenth thru twelfth grade enrolling an approximately equal number of students.[2]

The focus of our research was on the impact of this reform on the building of a sense of community among students and between students and teachers. We spent a good portion of our time in the school in interviewing approximately twenty-two of the school's eighty-two teachers, four counselors, the principal and assistant principal, and several district administrators. In addition, we held three focus groups with students and talked to four parents who had students enrolled in the school. A common theme that emerged from our interviews involved student-student and student-teacher relationships. One teacher noted in this regard that the bonds between students and teachers were "important" in creating a setting that was supportive of education. According to another teacher, smaller learning communities would make students "feel more in touch with school and the school environment." In describing the potential impact of smaller learning communities, teachers noted that students "feel like they are a part of something," and that they possess a "sense of belonging," All and all, in one teacher's words, the goal of this reform was "to make the kids feel a sense of community and to feel like they belong, to increase attendance, to make them feel connected to certain teachers because they will see them more often, to help them to focus on things that they are more interested in. . . ."

The school's administration saw the introduction of smaller learning communities similarly. The assistant principal commented that the "core reason that we did this was to improve the relationships, engage kids more because they had relationships with teachers." The principal also commented on the development of relationships. Students, he said, would from time to time meet with a counselor to ask for a change in a class. The response of counselors to such requests was to tell a student that such a shift would necessitate a "change in your learning community or your team." The typical student response, he went on to say, was "well leave it be. I don't want to change my team."

This was a viewpoint that some of the parents to whom we talked also held. One parent noted that smaller learning communities allowed student to "get to know a teacher a little better," which provided for a better connection between students and teachers. For another parent, this reform would "foster some sort of community feeling." Students were, as it turned out, far less aware of any changes in relationships that smaller learning communities had brought.

I had some twenty-four years earlier in 1986 written a book about education and community. Entitled *Building the American Community: The School Curriculum and the Search for Social Control*, the volume examined the efforts of educational reformers during the first half of the twentieth century to use the curriculum as a mechanism of regulation and control to construct what I referred to as a sense of community. This was, I noted, an attempt, quoting the sociologist Robert MacIver, to ensure "the like willing, or the like thinking, or the like feeling of social groups."[3] I do not think that the interviewees in the smaller learning communities study were as explicit and definitive when they invoked the concept of community as was MacIver or for that matter the educational reformers who I wrote about in the earlier book. Yet, I do think that these contemporary informants are talking about essentially the same thing.

The time that has elapsed since I wrote *Building the American Community* has provided me with much in the way of hindsight concerning the issues that the volume considered. There has been the appearance of the scene of an array of new, critical scholarship in education that has both refined and advanced our understanding of the regulative role that schools and the curriculum play.[4] My own research and writing have in the intervening period moved on to other subjects—most particularly the history of efforts to use the curriculum to enhance academic achievement and the interplay between curriculum and urban school reform—that I have explored using interpretive frameworks similar to the lens of community.[5]

Finally, there has been over the course of the almost three decades since I first wrote the book an increasing interest in the question of community that has spurred forward a virtual growth industry of research and commentary on this subject. There are as it turns out a number of interrelated factors that are responsible for this surge of interest. One has to do with the attempts of social critics to address the upsets and dislocations that they associate with the accelerating pace of those economic changes subsumed under the rubric of globalization. Another related factor involves the effort of politicians, scholars, and ordinary citizens to address a key feature of globalization, the changing role of the state from one of governing to that of enabling. It is a shift that requires a reassertion of the authority of civil society as well as a better understanding of how public private partnerships assume the tasks traditionally undertaken by the state acting alone. A third and final factor rests on the efforts of educators to promote civic engagement and citizenship as mechanisms to empower youth in the face of these social and economic changes. The concept of community has become, then, an interpretive lens for understanding these and related features of contemporary life.[6] My intent in this volume is to use that lens for interpreting instances of urban school reform since 1960.

II.

Community is clearly an alluring concept for one who wishes to understand such cultural practices as schooling. As was the case with those who I interviewed in conjunction with my smaller learning communities research, the term invokes an array of positive and desirable relationships including belongingness, common identity, consensus, intimacy, solidarity, shared values, and unity to name but a few.[7] In a world that is becoming increasingly complex as well as more dangerous, the concept of community conveys a feeling of safety and happiness.[8] And according to the sociologist Raymond Williams, it is a term that never appears to be used negatively, or is it pitted against a more positive state of affairs.[9]

The key feature of a notion of community in addressing these and other educational issues is its imprecise meaning, which allows proponents of an array of political and ideological persuasions to make use of it. As Alan Ehrenhalt notes:

> On the far left it is a code word for a more egalitarian society in which the oppressed of all colors are included and made the beneficiaries of a more generous social welfare system that commits far more than the current one does to education, social services, and eradication of poverty. On the far right it signifies an emphasis on individual self-discipline that would replace the welfare state with a private rebirth of personal responsibility. In the middle it seems to reflect a much simpler yearning for a network of comfortable, reliable relationships.[10]

Community, then, can be used in a variety of ways to refer to an array of different end states. The political scientist Robert Booth Fowler has identified five such viewpoints. There is the participatory community that is built on face-to-face relationships, self governance, and equality. There is also the republican community in which the ethos of the civic virtue and personal responsibility prevail. A respect for traditional values and a commitment to family, neighborhood, church, and nation also describe a form of community, which Fowler refers to as a community of roots. He goes on to locate a sense of community in the human desire for the survival of the planet that is built on environmentalism, sustainability, and peace. Finally, he points to the role that varieties of religious experience play in forming a sense of community.[11]

In this vein, the sociologist Nikolas Rose notes that notions of community have been invoked numerous times since the late nineteenth century, but their meanings differ. In the late nineteenth and early twentieth centuries, the idea of community referred to a sense of solidarity that would repair the dislocations caused by industrial capitalism and an accompanying division of labor. In the years after World War II the invocation of community involved effort to repair the kind of undermining of neighborhoods that had occurred as result of the rise of the impersonal, bureaucratic state and its agencies. In the 1960s community was a collective entity comprised of the various welfare institutions that were to be found in close proximity to where people lived and worked. And today when proponents of communitarianism speak of community, they are talking about an affective element that enters into the establishment of one's identity.[12]

Yet, there are problems with such a vague and imprecise notion that can undermine its attractiveness and perhaps its value. Virtually everyone seems to find a use for the concept. We have already seen its presence in popular discourse, in our case in the labels that teachers, parents, and students invoke to describe desirable relationships. We will in this chapter and later in the book find scholars making use of a notion of community to describe their differing versions of the good society. Even the nation's corporate sector finds something to value in talking about community. A brochure entitled "McDonald's and You" that is available at its restaurants throughout the United States proclaims that "we share one community" and describes its outlets as "socially responsible neighbors. The pamphlet is loaded with accounts of how the chain's legacy is one of "giving back" to "our communities," its support of programs that benefits "children, families and neighborhoods," and how McDonald's is acting "to preserve and protect our environmental resources for you."[13] Similarly, Starbucks distributes a brochure at their stores entitled "Starbucks in Our Communities." The flyer describes a range of efforts including contributing cash to nonprofit organizations in which company employees volunteer their time, partnering with international development and relief organization that address a variety of needs in coffee growing countries, and introducing energy conservation equipment in their roasting plants.[14]

A particularly egregious example of the corporate appropriation of the idea of community is a magazine by that very name published by Utah's Zions Bank. A glossy publication filled with pictures and articles on a range of subjects including prominent Utahans, health foods, estate planning, and seasonal recreational activities to name but a few the magazine is filled with an array of advertisements for furniture, jewelry, automobiles, vacation resorts, and other high end luxury goods and services.[15] The notion of community that the magazine conveys is certainly a positive and uplifting one. Yet, the community that it addresses is a decidedly small and restricted group whose common identity and shared values are so narrow as to exclude all but the most privileged and affluent.

Another problem involves the fact that there is something nebulous about the concept that leads those who use the term to define it in numerous, vague, and often time contradictory ways.[16] Suzanne Keller notes in this vein that the term has been used to describe a physical or geographical place, a set of shared values, the bonds and networks that join people and groups together, and collective entities of one sort or another.[17] In a 1955 essay, George Hillery identified 94 distinct definitions of the term in current sociological literature. While there were some common elements in these definitions, he did find much in the way of disagreements.[18]

A third problem associated with the concept of community has to do with its actual existence. The anthropologist Benedict Anderson notes in this vein the "imagined" quality of one particular kind of community, that of the nation. He argues that once the features of the notion of nationality or "nation-ness" as he calls it were spelled out in the late eighteenth century, they took on something of a life of their own as present in any entity given that designation notwithstanding its actual attributes.[19] A characteristic commonly associated with community is that of deep, face-to-face relationships among its members. For Anderson however, such relationships are not possible in anything beyond the smallest of such units. Yet, they are often posited to exist as one of the defining attributes of any community. They are in other words imagined.[20]

Derek Phillips sees the idea of community in similar terms. There is, he notes, a general agreement among those who theorize about community as to its meaning. Four characteristics are particularly salient, namely a shared locality, social interdependence, shared patterns of behavior, shared history, and a sense of belonging together.[21] The seventeenth century Massachusetts Bay colony is, according to Phillips, described as embodying these characteristics. Yet, the reality was far different. The colony did occupy a fixed territory, but it was one that was widely dispersed between a numbers of distinct settlements. It was hardly the village that a conventional view of community might imply. The inhabitants of the colony did share a common language and culture. Yet, they also embodied the decidedly anti-communitarian attributes of the adventurer and explorer, namely individualism, self-centeredness, and opportunism. The organization of the colony was not the kind of flat, roughly egalitarian structure that a notion of community might imply. Rather, it was organized hierarchically into different and unequal ranks from a ruling elite at the top downward to a diverse array of merchants, farmers, laborers, and servants.[22] The evidence from colonial America and elsewhere, according to Phillips, raises doubts that anything approaching a community has ever actually existed.[23]

Beyond these general problems with the notion of community, its use in an American context raises a particular difficulty. Americans have been and remain quite divided over what turns out to be the central question surrounding the issue of community, namely what constitutes the good life to which individuals might aspire. For much of our early history the answer to this questions has been found within a republican tradition that can be traced back to the debates of fifteenth century Florentine political theorists about virtue and corruption that made their way first to England and then to colonial America. According to this convention, the state played an important role in instilling individuals with civic virtue and a sense of public spiritedness that was required of citizens in a self-governing republic.[24]

This was a view of the state that seemed to fit the contours of rural America but in the midst of the mid-nineteenth century market revolution and the subsequent growth of industrial capitalism and urbanization gave way to a more voluntaristic outlook that replaced civic virtue with choice as the raison d'être for life in America.[25] Within this voluntaristic tradition, the state occupies an impartial position regarding any determination of what constitutes the good life. It offers a structure of procedural values including open mindedness, liberty, and fair play that allows individuals to engage in the kind of deliberations through which they can chart their own life destinies and their own sense of community.[26]

Because of these problems with the notion of community, it is not surprising to find commentators and critics voicing doubts about its efficacy and urging its elimination from our lexicon of possible societal visions. In an opinion piece in a recent issue of the British journal *Prospect*, William Davies notes that invoking notions of community has less to do with "revealing truths or referring to facts," than it has to do with "projecting a persona and shaping a situation." This is what he refers to as serving a "performative" function that rarely addresses or resolves important questions but rather reduces them to the level of distracting "moralisms."[27]

Yet, having said all of this, the notion of community seems to be an important one. In the *City of God*, Saint "Augustine invokes the concept to define the elements of "sound and just government" that are central to his criticisms of the Roman Republic. Without community, which he defined citing Cicero as "an association united by a common sense of right and a community of interests," Rome lacked the principles of justice that enabled it to be the commonwealth it believed itself to be.[28] Much closer to our own time, it was the vision of the "beloved community" that enlisted the ideals of redemption and reconciliation in Martin Luther King Jr.'s call for an integrated society free from racism, economic injustice, and war.[29] Embedded, then, in discussions of community are those most important of ideals, beliefs, and visions that bind us together and define for us a sense of the common good, that is the beliefs and other features of our culture that serve everyone but do not belong exclusively to anyone.[30] In a recent airing of the *Real Time* television show with Bill Maher, the mayor of Newark, New Jersey, Cory Booker made the same point when he spoke of community as that which brings people together "for purposes greater than ourselves."[31]

The concept of community plays a similar role in our educational discourse in pointing to our most important aspirations for public education.

In *Why Fly That Way? Linking Community and Academic Achievement,* Kathy Greeley, who teaches middle school in Cambridge, Massachusetts, describes her work with a group of eighth grade students in writing, producing, and performing a class play. The drama intersperses a parable that likens the cooperation evident in the V formation that characterizes the flight of migrating geese to the cooperative efforts of historical and contemporary characters involved in struggles for equality and racial justice. At the heart of the play and central to Greeley's book is a focus on the importance of a sense of community in creating a safe and supportive classroom environment that promotes academic achievement.[32] The book offers no one understanding of community. Rather, it suggests multiple meanings that point to such desirable goals of schooling as working for the common good, cooperating with one's colleagues, overlooking differences, and feeling sufficiently safe to reveal one's inner feelings and thoughts.[33]

Arnold Fege of the Public Education Network argues that the recent Federal No Child Left Behind Act has a broader purpose than just instituting an accountability regime into our educational practice. The data about the school performance of disadvantaged children that this legislation requires can provide parents with the kind of information that can empower them in their efforts to be advocates for the well being of their children.[34] To the degree that this act does this, he goes on to say, it fulfills the original democratic purpose of its parent legislation, the Elementary and Secondary Education Act, "of building community and civic ownership."[35]

We are left, then, with the dilemma of what to do with a concept that appears to be important and yet whose meaning seems so nebulous and indefinite as to lead many to doubt its usefulness. This is only a difficulty if we assume that the notion of community has a fixed meaning and that it refers to what Raymond Plant labels a "palpable object." within society.[36] "To regard all meaningful words as names, as having a wholly referential character," he goes on to say, "is mistaken. It involves on the one hand torturing some perfectly ordinary words upon a procrustean bed in order to make them yield meanings of the requisite logical type; on the other hand the meaning when so produced often does not do justice to the varieties of uses which the word may have in ordinary discourse."[37] We might, then, think of the notion of community as comprising multiple discourses—those textual ideas, concepts, and statements that not only provide meaning but constitute systems of power to affect social organization and human behavior—circulating in the conceptual space surrounding this idea. It is these discourses that represent the discontinuous and often conflicting pathways that shape, structure, and define the notion of community in the different ways that we have thus far discussed.[38]

One way in which we can overcome this seeming obstacle is to consider the notion of community as what is called a floating or empty signifier, that is as a word without a single or specific referent.[39] According to Rosa Burgos, a notion of this sort "flows and circulates throughout a variety of meanings, and sites of enunciation and it has become what it is today through a series of discursive articulations throughout history."[40] This is not to say that the concept of community has no meaning. Actually, I think those who use the term do so to point to connections between and among individuals and groups and their sense of collective belonging. Notwithstanding its multiple referents, it can be thought of as an indicator of those discursive practices that vary in time and place but serve the role of joining individuals to collectivities of various sorts.[41] My task in this volume, then, will be to consider what a conception of collective belonging or community tells us about urban school reform.

III.

Not surprisingly Americans have invoked the notion of community in decidedly different and conflicting ways. The term is often used within a conservative intellectual tradition to refer to the shared values and common world view associated with a specific geographic location and those who populate it. It suggests a social setting that is ordered and static and dominated by a ruling hierarchy. It is world view that is often employed to critique urban, industrial society. There is, however another usage, a liberal notion of community, that accepts the values of urban, industrial society, rejects locality as its key criterion in favor of what Plant refers to as "shared ends" emerging out of a "functional cooperation" among people. This is a viewpoint that rejects a yearning for the past in favor of an effort to reconcile the values of individuality with those of community.[42]

Those favoring a conservative notion of community have used the term in an exclusionary manner to point to those characteristics that separate individuals and groups from each other including differences in language, religion, ethnicity, and culture. Talking about community in this way serves first to limit membership to certain individuals and groups and once that has been done to constrain, isolate, and eliminate others. Although a feature of the notion of community that is viewed favorably by those who are within it and a part of it, it allows for its invocation to legitimate racism, sexism, and a host of other forms of control and exclusion.

Adherents of a liberal notion of community, however, have used the term in an inclusionary manner to emphasize the similarities among individuals and groups that bind them together. It is a notion of community that has been used to support efforts to expand political participation and cultivate democracy.[43] These two forms of community are sites that nurture very different kinds of social capital or interactive networks, social relationships, and connections. Exclusionary communities cultivate what Robert Putnam refers to as bonding types of social capital that in his words are "inward looking and tend to reinforce exclusive identities and homogeneous groups." Inclusionary communities, on the other hand, promote bridging forms of social capital that "are outward looking and encompass people across diverse social cleavages."[44]

A good example of this conflict appears in the way in which early twentieth century American intellectuals invoked notions of community when talking about increased immigration. There were those like Edward L. Thorndike, David Snedden, Ross L. Finney, and Edward A. Ross who saw these immigrants as representing a threat to social order and stability and ultimately to American democracy itself For them, the idea of community became something of a defensive notion designed to curb the social disruptions and dislocation that they identified with a growing diverse and heterogeneous population. Their solution of choice was immigration restriction. As a second line of defense they proposed using various social institutions including the schools as mechanisms of social control to instill in these immigrants what they perceived as correct values and attitudes. What they sought was a homogenous and likeminded community.

There were others, most notably John Dewey, Charles Horton Cooley, and George Herbert Mead, who took a far less defensive position. They did not fear immigrants but welcomed them as a source of fresh ideas and innovative practices that would enrich American society. Their understanding of community was one that was built on the mutual adjustment of all segments of society to a commonly agreed on set of values, attitudes, and standards of behavior reflecting the diverse cultural practices of the population. Securing this adjustment and mutuality was the task of a democratic brand of social control that was to be entrusted to a host of social institutions including the schools.[45]

There are as it turns out, different ways of measuring the presence of community, but none of them are all that precise. We have already seen that Robert Putnam, who defines community as the sense of belonging that we obtain from our membership in any of a number of social groups, gauges its presence by the degree to which the interactions and associations create social capital.[46] In his discussion of what he sees as a decline of community in post-1960s American society, Putnam identifies a host of indicators of diminished citizenship participation in those political, religious, workplace, and civic associations that serve as sites for the development of social capital.[47] He goes on to identify things that can be done to reverse this trend in declining participation, which in turn holds the potential for building social capital and restoring community.[48] More recently Putnam and two colleagues describe a diverse array of such efforts throughout the country, including the school reform initiatives of the Valley Interfaith coalition in Texas' Rio Grande Valley, the congregation building strategies at California's Saddleback Church and other so-called megachurches, the organizing tactics of the Harvard Union of Clerical and Technical Workers in Cambridge, and the development of a participatory work culture at United Parcel Service.[49]

Another approach at assessing the presence of community, particularly popular with political scientists studying urban renewal, is to associate community with the concept of civic capacity, which refers to the ability of various sectors of communities to join together to solve their problems. Clarence Stone and his colleagues note in this vein that:

> When city hall, business elites, and labor unions combine efforts to redevelop downtown or build a new convention center, a community's civic capacity has been activated. When a wide alliance develops enough of a common understanding to work in concert to reform urban education, civic capacity has been activated...[50]

The reference in this quotation to the physical redevelopment of cities is not happenstance. The notion of civic capacity can be traced to the work of political scientists interested in the role that partnerships between business and government, what they refer to as regimes, have played, beginning in the 1950s and continuing to the present day, in supporting such urban renewal efforts as attracting business to central cities, clearing slums, and promoting tourism. In recent years, many of these same scholars have extended this idea to include efforts at urban school reform.[51]

Civic capacity is in effect an assessment of the existence within urban settings of the conditions that allow for the formation of cross-sector partnerships that support systemic reform. Its activation requires that various interest groups within a community are able to transcend their individual concerns and become mobilized around a common understanding of the problem to be addressed. It also necessitates that a wide and representative array of community stakeholders participate in resolving the problem and that those stakeholders are willing to secure the resources required for such a resolution.[52]

Seen in this way, civic capacity bears some resemblance to the notion of social capital. Yet, there are important differences. Social capital refers to the norms of reciprocity, patterns of trust, and systems of networking that occur in interpersonal relations within families, volunteer groups, churches, and similar associations. Civic capacity, on the other hand, emerges within larger community relationships where disparate interests interact around issues of politics and governing. Continuing the kind of interactions and reciprocal relationships that build social capital serve to strengthen it. That is, the more people are able to interact successfully, the greater the resulting trust among them. This is not necessarily the case with civic capacity where efforts at collaboration may over time break down and result in a dissipation of trust and cooperation.[53]

The existence within social groupings of either social capital or civic capacity can certainly point to the existence of community. Yet assessing its presence can require us to make something akin to a moral judgment about the goodness or lack of goodness of social arrangements that are quite contentious. To avoid this problem, I want to use community not as an ethical assessment but as an analytic one to explore how different stakeholders involved in processes of school change pursued their efforts at reform. That these individuals and groups undertook different agendas for different purposes is not of particular importance. What is crucial is that using the lens of community to interpret such efforts helps us to understand their underlying intentions.[54]

Yet, we need to understand that the differing intentions of those who promote the building of community are not trivial matters. Jośe Rosario has noted in this vein how the efforts in two middle schools to divide themselves into smaller units that might have been inscribed with a communitarian "moral order" turned out to be structural reorganizations with little in the way of ethical content. What, in other words, might have been changes to create among staff, students, and administrators in these two schools a sense of collective belonging and common purpose became simple bureaucratic mechanisms that allowed their administrators to maintain order and centralized control.[55] Looking, in other words, at efforts at school reform from a conceptual lens of community is valuable in its own right. In doing so, however, we must be careful not to ignore what invoking a notion of community means in the way of actual practices.

Efforts to reform schools, particularly urban schools, are events that engender a host of issues and conflicts that have been interpreted through the conceptual lens of community. One set of such issues has to do with racial conflict. In the next two chapters, I will look at two instances where a notion of community was invoked in the context of school reforms designed to achieve racial integration and equal educational opportunity. Another set of issues involves school organization. The following three chapters will look at efforts at urban school reorganization in attempts to promote what might be thought of as a sense of community. The curriculum too is a site for struggles involving community. There is the issue of whose knowledge comprises the curriculum and whose knowledge is excluded. There is the related matter of who does and who does not get access to the knowledge, skills, and dispositions that the curriculum conveys. And finally there are an array of struggles surrounding the role that the curriculum plays in cultural reproduction and the perpetuation of patterns of power and privilege. Such issues involving curriculum will be found in all of the chapters in this book.

There are numerous examples of contemporary educational researchers and policy makers invoking notions of community to frame their arguments. Lea Hubbard, Hugh Mehan, and Mary Kay Stein's recent study of school reform in San Diego offers a good example of the role that a notion of community can play as an interpretive lens for understanding educational change. The focus of their research was the efforts of San Diego School Superintendent Alan Bersin between 1998 and 2002 to enhance academic achievement by reforming the district's instructional practices in reading through the introduction of what was called Balanced Literacy. A "literature-based" model for teaching reading, it had been introduced into New York City's Community School District # 2 in 1989 by its superintendent, Anthony Alvarado, as the key element in his reform agenda to improve reading instruction. Bersin hired Alvarado as chancellor of instruction in order to implement Balanced Literacy into San Diego's schools.[56]

The story of this reform that Hubbard and her colleagues tell is of a "centralized" and "fast paced" effort to make student achievement the central focus of the district and its personnel. It was an initiative that these researchers and others who studied it viewed as contentious and as one that resulted in ongoing patterns of conflict and hostility among the key stakeholders.[57] Yet, in interpreting the theory that they saw driving this attempt at change, they invoked a notion of community. As Hubbard and her colleagues saw it, the implementation of this reform was a learning process involving the various stakeholders. Taking their cues from proponents of situated cognition, they did not see this learning as something that occurred within individuals themselves. It was, they argued, a collective enterprise that:

> happens when individuals bring varying perspectives and levels of expertise to the work before them. As they work together toward shared goals, they create new forms of meaning and understanding. These new meanings and understanding do not exist as abstract structures in the individual participants' minds; rather they derive from and create the situated practice in which individuals are coparticipants.[58]

From this vantage point, these researchers viewed the school system as a collection of what they called "nested learning communities" in which school personnel in different positions at different levels of authority attempted to interpret the meaning of the reform, to identify and resolve inconsistencies, to learn from each other, and ultimately to act. In such settings, for example, teachers interacted with students to teach them academic content; principals interacted with teachers to assist them in improving student learning; central office personnel worked with principals toward developing their expertise; and district leaders related to a host of external constituents.[59]

Implementing Balanced Literacy, according to Hubbard and her associates, included two kinds of change. One involved a restructuring of the roles of the superintendent, principles, and teachers to make instruction their central concern. The other change, "reculturing," required:

> developing common thoughts and beliefs and a common language about skills, practices, and accountability among educators in every part of the school system to reach the district's goal of improving student achievement by supporting teaching and learning in the classroom. Reculturing also meant changing teachers' practices from actions conducted in the isolated privacy of their own classrooms to a public community of learners in which improved instructional practice was supported through professional development and leadership.[60]

Such a process required what they referred to as the "co-construction" of the reform in which there was mutual trust among the stakeholders and an ability for them to work collaboratively to attain common goals.[61] It was in effect a process that established a sense of collective belonging among the stakeholders.

A similar use of a notion of community has been embraced by the superintendent of the New Orleans Recovery School District, Paul Vallas. Having led school reform efforts in Chicago and Philadelphia, Vallas was appointed in 2007 to lead the Recovery District, one of three systems established in New Orleans to address the virtual breakdown of what was an already troubled school system in the wake of Hurricane Katrina. New Orleans schools, as Vallas sees them, faced a double assault. There was the impact of generations of "deep poverty" among the city's largely Black population that had produced the patterns of low achievement, student violence, and poor attendance that characterizes many urban school systems. "You compound that," he noted, "by the aftermath of the hurricane, by the physical, psychological, emotional damage inflicted by the hurricane," and you get a virtually non-functioning school system.

Reforming schools under these conditions, according to Vallas, requires more than enhancing the academic program. Schools have to become more like families and "provide the type of services you would normally expect to be provided at home." City schools, he goes on to say, would have to offer "three meals a day, including hot lunch and dinner" as well as "dental care and eye care." Vallas characterizes such schools by invoking the notion of community. "You begin to make the schools community centers.... The whole objective is to keep schools open through the dinner hour and keep schools open 11 months out of the year." What is called for, according to Vallas, is a sense of community that allows New Orleans to "close ranks" behind a restored school system.[62]

In the essays that he contributed to his edited volume *Inside Charter Schools: The Paradox of Radical Decentralization,* Bruce Fuller invokes the concept of community as his lens for understanding the contemporary charter school movement. As Fuller sees it, the impetus for this reform represents a modern variant of the same impulse that spurred forward the mid-nineteenth century common school movement. Both movements represented, according to Fuller, attempts to use schooling to instill within children a commitment to a set of shared values and beliefs associated with a notion of community.

Yet, the community that common school reformers yearned for was a different one from that which is being sought by proponents of charter schools. In the former instance, community became something of a centralizing notion that would unify the nation around a republican tradition of civic virtue, self-government, and social responsibility.[63] Proponents of charter schools, on the other hand, favor a narrower and decentralized brand of community. The unity that charter schools build is particularistic rather than universal and serves its supporters by "strengthening the borders of their own community."[64] The emergence of charter schools for Fuller represents something of a shift in the locus of authority from the central state to local entities.[65]

IV.

Since a central purpose of this book is to explore the interplay between curriculum and community, we ought to say something about that connection at the outset. Probably no individual has done more to cement that relationship than did John Dewey. His starting point was the rural, small town of nineteenth century America where the principal industrial and manufacturing activities of the day were centered in the home:

> The clothing worn was for the most part made in the house; the members of the household were usually familiar also with the shearing of the sheep,

> the carding and spinning of the wool, and the plying of the loom. Instead of pressing a button and flooding the house with electric light, the whole process of getting illumination was followed in its toilsome length, from the killing of the animal and the trying of fat to the making of wicks and dipping candles. The supply of flour, of lumber, of foods, or building materials, of household furniture, even of metal ware, of nails, hinges, hammers, etc., was produced in the immediate neighborhood.[66]

According to Dewey an array of social, economic, and demographic changes in the fifty years following the end of the Civil War and into the early days of the twentieth century led to the separation of the home from the workplace. These changes include the growth of corporate capitalism, the rise of the industrial plant, the movement of settlements from the countryside to the city, and the increasing diversity of the population in the wake of immigration.

Under these new conditions, the face-to-face relationships of the small town gave way to less intimate associations. The independent craftsmen, merchants, and farmers who owned and directed the products of their labor during the early days of the nation were transformed into employees and workers and linked by more distant and remote networks of interdependence. The once vital connections that told individuals who they were, what they did, and how they related to the larger society had become lost as experts assumed responsibility for decisions that were once under popular control. The result, Dewey argued, was the erosion of the ability of ordinary individuals to function as active citizens to resolve the dilemmas that they faced through participating in democratic politics. Their collective identity, which Dewey referred to as the "public" had become so dissipated and attenuated that it could no longer address the myriad of problems that they, their families, and their fellow citizens faced.[67]

The remote and distant interpersonal relationships brought about by industrialization and the growth of technology had produced, according to Dewey, a "Great Society." It was not, however, the "Great Community" with the robust "public" that he sought. What was missing was the interplay that would allow individuals to shape the groups to which they belonged while at the same time not limiting the potential of the group to manifest an overriding common purpose:

> From the standpoint of the individual, it consists in having a responsible share according to capacity in forming and directing the activities of the groups to which one belongs and in participating according to need in the values which the group sustains. From the standpoint of the groups, it demands liberation of the potentialities of members of a group in harmony with the interests and goods which are common.[68]

Dewey's vision of the "Great Community" was his challenge to American intellectuals like Walter Lippmann who in the years following World War I seemed to have lost their faith in the kind of democratic ideal that supported broad popular participation in political and social affairs. Ordinary citizens, according to Lipmann, lacked the time and interest to assemble the necessary information to make the array of informed decisions required for the smooth functioning of modern society. If they were to avoid being manipulated by elites, they needed to rely on an array of experts in various walks of life to form and shape their opinions.[69] While Dewey acknowledged Lipmann's work as representing "the most effective indictment of democracy as currently conceived ever penned," it was a view counter to his understanding of the direction that modern American society should follow.[70]

Early in his career while at the University of Michigan in 1888 Dewey had argued that the individual did not exist in isolation from others but was a "social being." What joined people together in society and what made democracy possible was the presence of a "common will" that would bring together majority and minority opinion into a unified whole.[71] For Dewey, this relationship between society and its members was a reciprocal one. As he put it, "if then, society and the individual are really organic to each other, then the individual is society concentrated. He is not merely its image or mirror. He is the localized manifestation of its life."[72]

What Dewey was driving at here becomes clearer in his 1896 essay entitled "The Reflex Arc Concept in Psychology." Not an essay on ethics or politics but on human behavior, Dewey challenged the prevailing psychological view of his day that used the metaphor of the arc to depict the relationship between an environmental stimulus and an individual's response. According to this view, the stimulus and response were separate and distinct entities. The stimulus provided a sensation that the individual reacted to with some sort of motor response. Dewey found this depicture of the relationship, the reflex arc in his words, to be inadequate:

> . . . it assumes sensory stimulus and motor response as distinct psychical existences, while in reality they are always inside a coordination and have their significance purely from the part played in maintaining or reconstituting the coordination; and (secondly) in assuming that the quale of experience which precedes the "motor" phase and that which succeeds it are two different states, instead of the last being the first reconstituted, the motor phase coming in only for the sake of such mediation.[73]

A circuit, Dewey believed, offered a better characterization of this relationship. The stimulus and response were not separate states but rather connected. As he put it, "the motor response determines the stimulus, just as truly as the sensory stimulus determines the movement."[74] As a psychological insight Dewey's critique of the reflex arc provided the intellectual grounding for the challenge that he himself, George Herbert Mead and other formative theorists of the discipline of social psychology would mount against John Watson's radical behaviorism and Edward Thorndike's connectionism.

At the same time, however, his critique of the reflex arc provided support for his social vision, specifically the reciprocal relationship he posited between individuals and society in many of his social writings including his ideas on community. It was as a community, Dewey argued, that the "public" could reassert itself.[75] Such an outcome required a recognition of shared goals to which all members of the "public" were committed.[76] For a community to exist, according to Dewey, "there must be values prized in common. Without them, any so-called social group, class, people, nation, tends to fall apart into molecules having but mechanically enforced connections with one another."[77] Yet, this commonness did not, according to Dewey, require homogeneity. Rather, what was called for was a bringing together of the various elements within the "public" to form something of a shared and mutual accord. A community was not a neutral entity that oversaw the competition among individuals each pursuing their own interests. It stood for something that was built out of the give and take and ultimate agreement between interacting parties.[78] What Dewey was talking about is what the sociologist Robert Bellah has referred to as "democratic communitarianism." Such a notion of community did not require total consensus. Yet, it demanded "something more than procedural agreement." There needs to be, he goes on to say, "some agreements about substance." The community that resulted:

> is one in which there is argument, even conflict, about the meaning of the shared values and goals, and certainly about how they will be actualized in everyday life. Community is not about silent consensus; it is a form of intelligent, reflective life, in which there is indeed consensus, but where the consensus can be challenged and changed—often gradually, sometimes radically—over time.[79]

Community, then, emerged out of the influence of both the group and the individual. Making virtually the same point more recently in a letter to an imaginary new teacher, high school junior Wilhemina Agbemakplido noted that "community makes you who you are, but I also make my community what it is."[80]

In a recent book, Lee Benson, Ira Harkavy, and John Puckett point to Dewey's 1902 speech to the annual meeting of the National Educational

Association for an illustration of the kind of community that Dewey envisioned. Entitling his remarks as "The School as Social Centre," Dewey called for the expansion of the school beyond its traditional role of teaching basic skills and conveying academic knowledge to a broader role of providing a foundation for a democratic society built from the interaction and participation of its members.[81] The solution to the problems facing modern society, Dewey believed, was the cultivation of "common sympathies and a common understanding."[82] The school:

> must provide at least part of that training which is necessary to keep the individual properly adjusted to a rapidly changing environment. It must interpret to him the intellectual and social meaning of the work in which he is engaged: that is, must reveal its relations to the life and work of the world. It must make up to him in part for the decay of dogmatic and fixed methods of social discipline. It must supply him compensation for the loss of reverence and the influence of authority. And finally, it must provide means for bringing people and their ideas and beliefs together, in such ways as will lessen friction and instability, and introduce deeper sympathy and wider understanding.[83]

The school and its curriculum were for Dewey key sites in the building of this sense of community. As Dewey saw it, the central educational problem of the day was the supposed conflict that educators and others claimed existed between the needs of society as embodied in the traditional academic disciplines and the interests of children. The problem for many of these educators was that those disciplines were not particularly appealing to children who were much more interested in the array of play and related activities in which they engaged outside of school. Dewey, however, believed that there was no conflict between school subjects and these non-school activities. They were actually compatible and reflected an underlying unity that Dewey believed joined them together.

To explain how these seemly different entities were one and the same thing, Dewey drew an analogy between an explorer's account of a discovery and the resulting map that a cartographer had drawn. Both depicted the same reality. Yet, their organization differed. The explorer's account represented the psychological organization that an individual gave to his or her trial and error wanderings across a landscape. The map presented the logical organization of the same territory prepared in accordance with the agreed upon rules and practices of cartography.

A similar relationship existed for Dewey between the interests of children and the traditional school subjects. Dewey identified the connection by posing the question of the origins of the academic disciplines. They first appeared on the scene, Dewey argued, as such day-to-day activities or as he called them "occupations" that early people undertook in order to survive. They included among other things cooking, measuring, cutting, planting, and building. Over the course of the development of civilization, these "occupations," which individuals organized themselves experientially or psychologically, were transformed into academic disciplines with their own logical structures. The mythological accounts that primitive people constructed to explain their origins were, according to this theory, transformed over time into history. And similarly, the efforts of, early people to plant, harvest, and prepare food provided the elements of what has become chemistry. What Dewey was in effect explaining was the development of human knowledge from its primitive organization in the day-to-day survival activities of people to its more mature form in the academic disciplines.

Based on this understanding of the origins of knowledge, Dewey saw the curriculum as the site where children would recapitulate the stages that occurred in its development. It was a mistake and did violence to the basic nature of children to introduce them to formal subjects as well as such skills as reading and writing too early and too suddenly. The curriculum in the earliest grades was to be composed of large and small group activities based on everyday human occupations that would gradually prepare children for the study of traditional school subjects. Cooking, for example, was to be used to introduce children to the basic elements of chemistry. Involving children in sewing would set the stage for studying the history of clothing, which in turn could lead to an examination of the transformation of fibers into fabrics through weaving and spinning. These were activities that could ultimately lead to the study of science and even geography. Similarly, carpentry would require the study of measuring and calculating that prepared children for the formal study of mathematics. All of these activities involved children in the use of written and oral language, thereby becoming venues for the teaching of reading and writing.

The curriculum that Dewey favored did more than involve children in the study of the traditional academic disciplines. The school, he argued, was more than a place for conveying certain knowledge to children and training them in certain habits. Organized around real life activities and projects that emerged from student-student and student-teacher planning, the classroom and school as Dewey envisioned them were the components of a miniature society. This was to be a place where students would learn and practice the skills, attitudes, and dispositions of participation, interaction, and problem solving that were the perquisites for democratic citizenship. These were the attributes that were required on the part of interacting individuals and groups for a sense of community to exist and for the "public" to reassert itself.[84]

Despite holding to a theory of community, Dewey, according to Benson and his colleagues, did not tell us what that community might look like or how we would get there. They note that the first President of the University of Chicago, William Rainey Harper, was involved in turn of the century efforts to improve the city's public schools. His appointment of Dewey to head the Department of Pedagogy as well as the Department of Philosophy at Chicago was designed to strengthen the university's work in the area of education and to connect the university with the work of school reform.

There was, however, a reform effort that came on the scene at the end of the nineteenth century and the beginning of the twentieth that sought to use the schools as well as other institutions to improve community life.[85] There were, Benson and his colleagues point out, proponents of this use of the schools who during the 1930s invoked ideas of community that they explicitly attributed to Dewey to address actual educational problems.[86] They point in this vein to the work of Elsie Clapp in Arthurdale, West Virginia and Leonard Covello at Benjamin Franklin High School in New York City to organize the schools as agencies for both exploring local problems in such areas as community health, parent education, recreation, housing, and sanitation within the curriculum as well bringing together students, educators, and community members to attempt to resolve them. The schools would as a consequence become venues for bringing together diverse viewpoints and reaching common understandings and solutions out of which a sense of community would emerge.[87]

Covello's work at Benjamin Franklin High School represented a good example of the attempt to connect the school with its community. As he put it, "the school must enlarge its vision, expand its facilities, and reach forth into community life in order to establish a magnetically intimate contact with the people, their problems, their potential values, and their needs."[88] The extent to which he achieved this goal, however, was limited. There were a number of community programs that operated both inside and outside of the school that established this linkage. One involved securing citizen participation through the establishment of a Community Advisory Council (CAC), agencies known as "street units," and a number of federally funded Works Progress Administration (WPA) initiatives. The CAC was composed, of a number of committees that addressed such community problems as health, citizenship, parent education, race, and guidance. The "street units" were agencies that were physically located outside the school in storefronts in the surrounding neighborhood that brought community members, business persons, parents, teachers, and students together to improve community life. These units included the Association of Parents, Teachers, and Friends whose task was to expand the work of the school into the community, the Friends and Neighbors Club that among other things supported a "friendship" garden and provided meeting space for community groups, and the Italo-American Educational Bureau and Hispano-American Education Bureau that developed educational programs as well as provided assistance for Italian and Spanish speaking immigrants for getting jobs and gaining citizenship. The WPA funded programs included support for a number of initiatives including an array of community oriented research projects, an adult evening and summer school that offered classes in English and citizenship for immigrants, and a remedial reading clinic.[89]

The regular high school program, that included both academic and vocational programs, was less successful in achieving this connection. Although efforts were made from time to time to integrate content about critical community issues into regular courses, there was no attempt to create a problem-centered curriculum. It remained for the most part a "teacher and subject-centered" one. There was in fact only one course at the high school, a leadership course for high ability students, that was organized around contemporary problems and involved the students in actual community research.[90]

Writing during the 1940s Ralph Tyler offered an example, not unlike the work of Covello in New York City, of how the schools, in Holtville, Alabama, were linking their work to the problems of the community. One such dilemma, he noted, was the declining productivity of the region's farmers. From their reading and study, the students identified a number of ways of increasing yields, such as crop rotation and diversified farming, techniques they in turn communicated to local farmers. They also learned how check dams might be used to control soil erosion, which was also causing declining productivity, and actively participated in programs for constructing them throughout the area.

Another problem that Tyler described from Holtville's community school core curriculum was that of nutrition. After studying the diet of local residents, the students suggested a number of ways that farmers could improve the nutrition of local residents, including the production of more diversified crops and the establishment of an area refrigeration facility so that fresh produce and meat could be readily available throughout the year.[91]

Currently Benson and Harkavy, who are faculty members at the University of Pennsylvania, are involved in an initiative to apply Dewey's thinking about community to school reform in the West Philadelphia neighborhood surrounding the university. Their vehicle for this effort is the West Philadelphia Improvement Corps (WEPIC), a partnership between the university, the School District of Philadelphia, and the local community. WEPIC is one phase of a larger effort on the part of the university to rejuvenate the community that involves the development of partnerships to improve local schools, build affordable housing, enhance economic development, expand city services, and create safer neighborhoods.[92]

A central feature of the WEPIC initiative is the provision of university support to create a number of community schools for West Philadelphia. The university provides significant financial support for establishing and maintaining these schools as well as involving university faculty, students and staff in all aspects of their operation including the selection of the curriculum, the teaching of classes, student mentoring, professional development, and the offering of special programs for students and for community residents. As part of its school improvement plan, the university created the Center for Community Partnerships that among other efforts supports the development of service learning courses in which university faculty and students work on experiential projects with students in these neighborhood schools.[93]

In one such effort, a university course in nutritional anthropology was transformed into a seminar on the relationship between obesity and nutrition that was taught by university students at a neighborhood middle school. The course was designed to enhance student knowledge about nutrition as well as to change their eating habits. As part of the seminar, university, and middle school students worked together on evaluating the weight, height, and body mass of middle school students; evaluating their diets; interviewing community families regarding their children's nutrition; observing students eating habits in the school cafeteria; and determining the sites where various kinds of food were offered in the neighborhood surrounding the school. The project also involved the middle school students in assisting university anthropologists in collecting, organizing, and interpreting nutritional data that they were collecting in the neighborhood. Other similar service learning courses included a comparative study of Philadelphia and fifth century BC Athens to explore the relationship between community, neighborhood, and family and a course involving university and neighborhood elementary and secondary students in the study of the nature of American identity. These courses were in effect collaborative endeavors to advance the knowledge and learning of both university and public schools students as well as improve the West Philadelphia neighborhood.[94] This effort like the work of both Clapp and Covello during the 1930s represents an attempt to use the schools as forums to bring individuals together to establish a sense of collective belonging around common concerns or issues.

It is the kind of community that Dewey and other like-minded individuals envisioned that will serve as our lens in the remaining chapters of this book to interpret post-1960 urban school reform. Such an approach will prove particularly useful as we explore the interplay between school reform and the educational aspiration of urban African Americans in chapters two, three, and four. In that vein, religious studies professor Eddie Glaude, Jr. has embraced a Deweyan brand of pragmatism to interpret what he calls the period of "post-soul politics" following the civil right and Black Power movements. He argues that there is today all sorts of evidence pointing to the success of African Americans including the growth of a Black middle class and increasing Black access to higher education and political power. At the same time, he notes, there is a growing Black underclass, an expanding incarceration rate among Black males, and an ever growing population of economically distressed single parent families. It is a contradiction best seen in the rising prominence of Black Americans like Colin Powell and Condeleezza Rice while at the same time witnessing devastating impact of Hurricane Katrina on much of New Orleans' Black population.

Glaude raises the question of how African Americans should respond to the dilemmas that they currently face. On that score, he rejects the highly racialized politics of the 1960s and 1970s that emphasized Black identity and solidarity to challenge racial segregation and promote civil rights. Contemporary African Americans, he goes on to say, are too quick to embrace the language of Black Nationalism and the tactics of mass protests to combat the problems that they face today. Doing so, he notes, falsely assumes that this ideology is as relevant to the present moment as it was in the past. It also mistakenly assumes that Black Nationalism is a fixed point of view with agreed upon characteristics that apply across a number of different times and place.[95]

What is called for on the part of contemporary African Americas, according to Glaude, is a more democratic and participatory political strategy that seeks a Black public much like the public that Dewey sought. The civil rights and Black power movements, Glaude argues, were two of several mass movements of African Americans during the twentieth century that in Dewey's terms mobilized an eclipsed Black public into a "Great Community" of broad and democratic participation to challenge and defeat the particular forms that racism had taken in the three decades following World War II. The shape that such a Black political movement will take, Glaude notes, is hard to predict independent of the specific problems of the moment that it seeks to address.[96] Yet, it will no doubt be a community built out of the shared goals, give-and-take, and mutual adjustment of ordinary citizens that Dewey so valued.

V.

The remainder of this book comprises a number of case studies of urban school reform occurring since 1960 that I will examine using the idea of community as a conceptual framework. In chapter two, I will explore two 1960s reform movements in New York City designed to enhance the school success of Black and Puerto Rican youth. They include the More Effective Schools initiative and the Clinic for Learning. I next move in chapter three to Detroit and consider the events surrounding the 1966 Northern High School student walkout. A response of Black Detroiters to their dissatisfaction with the largely white administrators and teachers who ran the city's schools, the boycott signaled their intent to secure their control over the schools and to launch their own reform agenda. Chapter four continues my study of Detroit but jumps ahead to the end of the century to examine the state initiated and sponsored mayoral takeover of the board of education. Chapter five looks at the efforts of urban educational reformers during the 1990s and into the twenty-first century to introduce partnerships as a strategy for improving urban schools. My focus will be on initiatives in three cities, New York, Detroit, and Minneapolis, At first glance chapter six may seem like a departure from the account I have developed thus far because I jump across the Atlantic to Great Britain to consider an early educational reform of that country's New Labour government, Education Action Zones (EAZ). This, however, is not the case. Partnerships as they have appeared in recent years are encased in an ideology about the economic and demographic transformations associated with globalization that are not self-evident or even very explicit in their American context. Yet, their ideological assumptions become clearer in this British reform initiative. So although the setting is decidedly different, it will provide a better opportunity than the American case studies of the previous chapter to understand partnerships from the vantage point of a notion of community. In chapter seven, my research assistant and I explore the transformation of a comprehensive high school into a number of smaller learning communities and consider what happens when those involved in this reform explicitly set out to build community. Introduced as a vehicle for both creating a more supportive and caring environment and enhancing academic achievement, we will explore how these two goals played themselves out in practice during a year in the life of this school. This chapter completes my account of school reform since 1960 by in a sense concretizing it in the context of an ongoing effort at reform in an actual school setting. I will conclude this volume with an epilogue in which I consider how we can recalibrate the idea of community that I developed heretofore to the demands of a twenty-first century globalized world.

This book can be read in two different ways. It is a collection of distinct case studies that pursue a similar theme. Chapters one through four are historical case studies, while chapters five through seven are ethnographic or perhaps more aptly ethno-historical.[97] They can consequently be read separately and in any order. The book can also be read as a single account. Although the chapters are tied together temporally and except for chapter six by place, they do not constitute a traditional historical narrative that depicts a succession of related events propelled over time that are moving according to some plan toward an ultimate goal. Michel Foucault's notion of genealogy is a more apt description of this volume. According to Mitchell Dean, "genealogy is a way of linking historical contents into organized and ordered trajectories that are neither the simple unfolding of their origins nor the necessary realization of their ends. It is a way of analyzing multiple open-ended, heterogeneous trajectories of discourses, practices, and events and of establishing their patterned relationships, without recourse to regimes of truth that claim pseudo-naturalistic laws or global necessities."[98] Foucault in fact referred to his genealogical approach as a history of the present to convey the point that his exploration was not to explain the past in terms of the present.[99] Rather, genealogy was a means of using our historical understanding of past cultural practices to disrupt what we often take for granted about the present, which in turn allows us to understand the present by problematizing those current practices.[100]

Although we begin our account around 1960, there is actually no distinct point of origin for the development of the idea of community. There are rather numerous discontinuous pathways, sometimes parallel and sometimes overlapping, that follow different lines of descent that constitute the multiple meanings that we give to a notion of community in the present. The case studies that we will examine in this book constitute distinct moments in time in the histories of these pathways. Looked at together they give us something of an account of the emerging arena of school reform. It is not, however, a progressive or evolutionary picture but rather something more haphazard written from the perspective of the present.[101]

Read either way, the book is an example of what Frank Fisher refers to as a policy narrative. Each case study tells the story of a dilemma and how that dilemma is addressed through public policy, in these cases educational policy. As in any story there is a beginning in which the problem is identified, a middle that suggests a policy solution, and an ending that explores the consequences of that policy in action.[102] Framing these dilemmas as narratives gives something of a human face to the often dry world of policy deliberation. They enable us to encase these deliberations within a plot involving actual people and events, conflict, and contingent results. They make clear that there are numerous policy outcomes to any dilemma that hinge on the context in which the problem arose as well as on the cast of characters involved and their beliefs and predilections. Not only do they depict what happens in the real world when people set about to resolve their problems, they serve policy makers as clear reminders that their work has real consequences for real people.[103]

Chapter 2

Community Conflict and Compensatory Education in New York City: More Effective Schools and the Clinic for Learning

Writing in 1968, David Cohen, who at the time was on the staffs of the Metropolitan Applied Research Center and the Harvard-MIT Joint Center for Urban Studies, noted the potential role that a major compensatory education program in New York City, More Effective Schools, could play in joining together the interests of New York City's teachers and their Black and Puerto Rican clientele. The key features of the MES Program, including smaller class size, reduced student-teaching ratios, and increased clerical support, would, Cohen noted, certainly improve the working conditions of teachers within the schools. They would also create a learning environment that would promote the academic success of children attending these schools. Suggesting this common purpose, Cohen borrowed the slogan that the city's teachers union, the United Federation of Teachers (UFT), had invoked in its strike the previous year, "Children Want What Teachers Need," as the title for his article. In effect, then, the MES Program seemed to place the interests of teachers in line with those of the city's Black and Puerto Rican communities.[1]

This sense of a common purpose did not, however, Cohen argued, really develop. A number of evaluations of the MES Program, which we will consider in this chapter, had raised doubts about the efficacy of compensatory programs, at least at the level that they were then being funded in the city's More Effective Schools, to enhance student academic achievement. The evaluations also, according to Cohen, lent credence to an emerging view among the city's Black population that the UFT was really not interested in Black schools and the success of Black students and going further that the largely White membership of the union was at bottom racist.[2]

It might be the case, Cohen noted, that increased expenditures would provide a stronger MES Program. Yet, the financial realities facing New York City would make it unlikely that additional city money would be forthcoming. He went on to argue that funding realities coupled with the meager impact on student achievement that evaluations of the program had noted lent support to the growing viewpoint among urban Blacks that the low achievement of African American children was the direct result of their lack of control over the schools that their children attended. As increasing numbers of Blacks saw it, according to Cohen, the only viable solution was decentralization or community control.[3]

Like many New Yorkers during the 1960s, Cohen was using the terms decentralization and community control interchangeably to refer to reforms that were designed to shift authority for administering schools from a citywide board of education to boards within much smaller subunits of the city.[4] He felt that such arrangements, which he labeled as "separatist," were the wrong answer. If it was difficult to obtain adequate support for programs like More Effective Schools under present conditions, community control would only worsen matters. White New Yorkers, he felt, would not be willing to provide support to schools that served only Black and Puerto Rican students. If minority children were to succeed academically, Cohen argued, they needed to be educated in the same setting with White children. Toward that end, he advocated the creation of large educational parks that were centrally located and served children from diverse ethnic and class backgrounds.[5] Cohen was in effect advocating the introduction of further school centralization at a time when the prevailing climate of opinion was in the direction of decentralization.

Recognizing the financial limitations that stood in the way of providing additional resources to programs like MES, Cohen concluded his essay by suggesting that it might be necessary for schools and teacher unions to establish partnerships with universities and other agencies to adequately fund the educational programs and services that disadvantaged students required.[6] Such collaborations would occur as we shall see later in this chapter and in subsequent chapters. It was, however, the call for community control that would dominate educational debates in urban schools during the 1960s. Such control might as its New York City proponents claimed unite the city's Blacks and Puerto Ricans around a single educational vision. Yet for Cohen, it was a strategy that would separate them from the city's White population and any hope of establishing a shared vision of the common good.

In voicing his concerns about More Effective Schools, Cohen was entering into an ongoing dispute about the very meaning of the concept of community. There were those who saw the MES Program as a vehicle for providing ghetto schools with the additional resources and services that would make them attractive to experienced, largely white teachers and to white middle class students. From their vantage point, this was an initiative that would promote integration. On the other side, he noted, were those who "racialized" the notion of community and made it a "synonym for black." They were advocates of community control, which, as he saw it, rejected integration in favor of placing the governance of schools serving African American children under the control of Blacks.[7]

Two years earlier in November of 1966, Rose Shapiro, a member of the New York City Board of Education and a community activist, reported to the board about a meeting that she and school Superintendent Bernard Donovan had attended earlier that month with about seventy African American and Puerto Rican parents concerning their dissatisfactions with the schools that were purportedly serving their children. At the meeting, these parents noted an array of problems that they saw as indicative of the inadequate education that their children were receiving. At P.S. 158 in Brooklyn, they reported that overcrowding had resulted in half-day sessions for first through fifth graders. Parents of children enrolled in Manhattan's P.S. 194 complained that their children rarely were given homework and when they were, teachers never corrected it. At P.S. 123 in Manhattan, children were not allowed to take their textbooks home because there was not a sufficient number for the school's population. Children requiring remedial services at Brooklyn's P.S. 76, according to parents at that school, were placed on a waiting list. And at P.S. 133 in Manhattan, a teacher was reported as having publicly stated about one of her students that "this child has no brains."[8]

These complaints are not unique to New York City or to the decade of the 1960s. They are examples of what has been and remains a national movement on the part of economically disadvantaged urban dwellers, particularly Blacks and Latino/as to challenge the quality of education that they believed those who run the nation's big city school systems are providing their children.[9] There is, however, more to be seen here. And that has to do with how since the 1960s a notion of community provides a lens for interpreting and explaining efforts at reforming urban schools. Each in their own way, this chapter and those that follow will explore this issue.

The focus of this chapter is New York City during the 1960s. My account considers two educational reforms. At the beginning of the decade, the United Federation of Teachers (UFT), which had just become the sole bargaining agent for the city's teachers, proposed its More Effective Schools Program as a way of improving the quality of education in inner city schools. Six years later, New York University entered into a partnership with the city's board of education to establish another program to improve education in the inner city, its Clinic for Learning at Whitelaw Reid Junior High School (Junior High School # 57). In this chapter, I will use the concept of community that I have previously developed to examine these two programs.

A good deal has already been written about the educational conflicts surrounding schooling in New York City during the 1960s. Much of this research has been directed toward the events preceding the now famous 1968 teacher's strike, the strike itself, its aftermath, and the controversy over decentralization and community control.[10] Understandably, these accounts focus most of their attention on workplace issues involving teachers, their union, and the board of education and on the racial conflict that divided teachers from the city's Black and Puerto Rican parents and community organizations. Issues of curriculum and pedagogy, teaching, and learning, which are the topics of this book, are certainly not ignored in these accounts. Yet, they are not considered as central issues in their own right but rather as venues for exploring larger battles about working conditions, the rights of teachers as workers, and race. The story that I will tell in this chapter and in the book itself is a different one. Its focus is on the about the classroom and the programs and policies that have been introduced there that address the question of building community and establishing a sense of the common good.

The More Effective Schools Program and the Clinic for Learning represent two examples of a larger national effort ostensibly designed to improve the academic achievement of disadvantaged minority youth. In 1961, the Ford Foundation embarked on what would become its major effort during the decade to address the problems of poverty, blight, and decay that had affected the nation's central cities. Known as the Gray Areas Program to refer to the condition of these inner city neighborhoods, this initiative included projects to improve urban governance, to combat juvenile delinquency, to enhance the delivery of social services to the urban poor, and to improve the quality of city schools.[11] One phase of the effort, the Great Cities School Program, supported an array of programs in ten cities designed to address the problems of urban schools and their students. In Detroit, Ford money provided for the adaptation of reading materials to fit the interests and abilities of students, the use of remedial or "coaching" teachers to provide additional reading instruction, and the establishment of an ungraded class to provide both remedial and advanced instruction for children who were performing below or above their grade level. As part of their efforts, Baltimore school officials developed a series of new tests to assess the abilities of disadvantaged children. And Milwaukee created a number of "orientation centers" throughout the city to place disadvantaged children in schools that were most appropriate for them. Over the course of the decade, the program would expand to include additional cities and a host of different strategies.[12]

This was clearly not the first time that schools sought to address the achievement problems facing their students. Programs for low-achieving children have been around in one form or another since the mid-nineteenth century. They underwent a rapid expansion in the years between 1880 and 1930 to include not only special and remedial education to address the academic needs of students but also medical inspection, health services and education, visiting teachers, vocational guidance, and sex education to provide for an ever increasing number of school related problems facing a growing and diverse school population. By the 1960s, programs, such as those funded by the Ford Foundation, under the rubric of compensatory education, would be added to this array of services to provide for the academic and social problems of urban, minority youth.[13]

II.

In 1963, the United Federation of Teachers, which two years earlier had become the sole bargaining agent for the city's teachers, challenged a proposal then being advanced by the New York City Board of Education to address academic problems in its most difficult schools. The plan, which was to be allocated some two million dollars, would pay approximately two thousand experienced teachers an annual bonus of $1,000 if they would be willing to transfer to so-called Special Service Schools, schools located in inner city areas and serving large numbers of low-achieving children. For the UFT, this proposal represented what they referred to as "combat pay" that would not redress the problems that made these schools undesirable placements for teachers. They went on to claim that the plan would actually make the situation in these schools worse by undermining the morale and commitment of those teachers who already worked there. The UFT suggested that the board use the money to bring conditions up to what they called "teachable" levels in ten Special Service Schools, and in return the union would encourage teachers to transfer to these schools.[14]

Initially the board rejected this counter offer on the grounds if it was successful, every school in the city would want similar provisions thereby costing the city billions of dollars that it did not have. Not dissuaded, the union initiated a campaign to promote its plan among the public and by so doing to convince the board to embrace it. The board, seeing that the plan had wide public support, came up with its own alternative, an after school tutoring program. Operating four afternoons during the school week and two hours on Saturday morning, the service would be staffed by teachers within the participating schools who would be paid almost twelve dollars for each two hour session that they conducted. Students would attend three tutoring sessions a week where they would receive remedial assistance and help in completing their homework assignments.

The UFT criticized the proposed program on several grounds. They noted that the plan had been developed by the board without consultation with teachers, parents, the community, and the children themselves. Such an approach, they went on to say, violated the spirit of the contract that the union had set up with the board that saw their relationship as a partnership. They questioned the efficacy of an after-school program that was limited to remedial work and supervising students in the completion of their homework. They felt that such an effort had to include an educational component during the regular school day. The union had doubts about the proposal to pay teachers a stipend for working in this tutoring program. Such a payment, according to the UFT, was an insult to teachers that pointed to the failure of the schools to properly educate their students. And notwithstanding the desirability of providing teachers stipends, the union felt that the way it was done was unfair. Teachers who worked in other board sponsored after school programs were prohibited from participating in this initiative. And these projects paid $7.50 an hour compared to the almost twelve dollars stipend provided to teachers in this proposed tutoring program. The UFT raised other criticisms including the absence of selection and screening procedures for children and teachers, the lack of an orientation and training program for teachers and other personnel, and the lack of an evaluation plan.[15]

Almost a year earlier, an ad hoc UFT committee had begun to meet to develop its own plan for addressing the problems of the city's schools. What they proposed was a complete school program conducted during the regular school day in small classes of around fifteen to twenty-two students, in schools with enrollments of between 800 to 1,000 students. To ensure that these schools were led and staffed effectively, the committee recommended a special screening process for selecting principals and teachers. They noted the need for a budget that would enable the schools to provide the extra services and materials required in the program. The committee recommended that these schools cooperate with their surrounding communities and with parents in carrying out their work. They called for an on-going evaluation plan and for the provision of up to date instructional tools. Finally, the committee supported efforts to ensure that the schools in this program were racially integrated.[16]

Racial integration in New York City was, however, contentious. In 1954, the board of education passed a resolution in support of the Brown decision. Yet unwilling to challenge the resistance of many White New Yorkers to busing and racial balance schemes that compelled White children to attend schools in minority neighborhoods, the board plans were typically incomplete, half hearted, and failed to bring about an integrated school system. The UFT had a strong commitment to civil rights and equal educational opportunity. Yet, it was a belief rooted in the race blind and meritocratic outlook of American liberalism and the labor movement. As a result, the UFT often hedged when it came to the question of integration, especially if support meant that large numbers of White teachers ended up being transferred to inner city, ghetto schools.[17]

The UFT spent most of 1963 in discussions with the board concerning the shape of this proposed program, responding to counter proposals from the board, and in building support among New Yorkers generally.[18] The latter issue posed dilemmas for the union as it sought to convert skeptics to its approach. They had, for example, to sell the plan to parents whose children would end up being transferred out of participating schools to allow for the class and total school enrollments upon which their plan was predicated.[19] To help garner that support, the union established a Citizens' Committee for Effective Schools headed by civil rights leader and AFL-CIO vice president, A. Philip Randolph.[20]

In April of 1964, School Superintendent Calvin Gross agreed to use the UFT proposal as the starting point for discussions to finalize a plan. He asked the UFT and the Council of Supervisory Associations to each appoint four representatives to meet with four members of his staff to draft a proposal. Over the next two weeks, this group, which was known as the Joint Planning Committee, met daily and ultimately came up with a plan that was approved by the city's board of education and introduced to ten city elementary schools in September and known as the More Effective Schools program. The UFT had originally called its proposal the Effective Schools Program but changed the name to offset the impression that non participating city schools were ineffective.[21] Not all that different from the proposal drafted by the UFT committee, the program included the recommended classes of twenty-two students, heterogeneous grouping, lengthening the time that the school was open from 3 p.m. until 6 p.m., the recruitment of committed teachers, and the provision for cluster teachers for each group of three classes to assist regularly assigned teachers by providing them with a daily preparation period.[22] Ultimately, it would be the

More Effective Schools Program and not integration that would come to represent the principal strategy that the UFT embraced in its battle for educational opportunity for the city's Black and other minority youth.[23]

During the first year of the program, children attending the ten MES schools exhibited major gains in reading performance. In a report to the board of education in November of 1965, Superintendent Bernard Donovan noted that prior to the inauguration of the MES Program students in these schools enjoyed six months of gains in their reading performance over the course of eight months. With the beginning of the MES Program, these same students showed achievement increases over the course of eight months ranging from seven to twelve months.[24]

An August 1966 evaluation of the program conducted by the Center for Urban Education suggested that the program was having positive results. Testing data, although not fully reported by the time of the evaluation, indicated gains in reading scores for third and fifth grade students enrolled in participating schools. Teachers were enthusiastic about the program and their morale was high. Yet, the observation team did note some concerns. There were indications, they noted, that there was some confusion as to the duties of cluster teachers. Teachers, they went on to say, were ambivalent about the heterogeneous grouping of students and in fact called for homogeneous grouping in mathematics and other skill oriented subjects. The observation team felt that the value of heterogeneous grouping to the program's success needed more justification than it had up to that point received and that teachers needed help in implementing this practice.

The evaluators also voiced concerns about what actually occurred in MES schools. They felt that the curriculum and teaching methods were not all that different from the practices of other non-MES schools, that few innovative or experimental practices had been implemented in participating schools, that teaching practices seemed more appropriate for large classes than for the twenty-two student class size mandated by the MES Program.[25] They concluded their comments on this issue by noting that "there seemed to be little in the curriculum that was helping the school to reach the disadvantaged student."[26] Among their recommendations were the needs for more in-service training for those working in MES schools, ongoing evaluation of the program, the extension of More Effective Schools to the junior high level, and increased funding.[27]

That same month, the board of education conducted its own evaluation of the More Effective School program. More favorable than the Center's assessment, this evaluation noted that MES schools had implemented, at least partially, if not fully all of the components of the program as it was conceived. One area in which the program fell short was that of racial integration. According to the evaluators, less than half of the MES schools were integrated. The report also noted that the academic performance of students enrolled in MES schools in reading and arithmetic improved over the period in which they were enrolled in the program and the gains made were greater than expected. The evaluation did single out program costs for comment. The evaluators noted that the cost per student for those enrolled in the MES Program was between \$860 and \$930, which was roughly double the cost per student in other city elementary schools.[28]

The union and the board of education read these initial results quite differently. The UFT called for the addition of 20 additional schools to the program while at the same time sending representatives to other union locals around the country to promote the introduction of the MES Program into their schools. As Superintendent Donovan saw it, the program was not particularly effective. He noted that although it cost the city twice as much to educate a child in an MES school, the actual public benefit was far less. UFT President Al Shanker responded by noting that the MES Program was the only effort then going on that had increased the reading scores of children in inner city schools to the level of national norms. The board of education, he went on to say, was more interested in saving money than in educating children.[29] This conflict pointed, according to Shanker, to the hostility between the board of education and the UFT. Shanker claimed that Superintendent Donovan did not like the MES Program from the beginning because it worked. Similarly, he asserted that the board of education disliked More Effective Schools because coming as it did in the early days of the union, it made the UFT look good in the eyes of the public.[30]

Throughout 1967 a war of words took place over the status of the MES schools. Supporters of the MES Program held a rally at the beginning of February at the board of education to demand full support and expansion of the program. Si Beagle, Chair of the union's MES Committee, noted that unless proponents of the program mounted a sustained effort to support it, the superintendent with the board's support would replace it with a less costly program. What they had in mind was up to twenty so-called "experimental" elementary schools that would combine elements of the existing MES Program with the All Day Neighborhood Schools (ADNS). The ADNS Program had been established in 1936 under the sponsorship of the city's Public Education Association and sought to use the school as a center for providing educational, recreational, and social service programs for disadvantaged children. The resulting program, according to Board of Education President Lloyd Garrison would be "almost as good" as the existing MES Program. The UFT, however, showed no interest in this plan, and the proposal went no where. Beagle had hoped that the school administration, the board, and other supporters of the MES Program would join together to petition the Federal government for funds to expand the program. Rather what was coming, he feared, was a battle between the school district and the UFT over the continuation of the MES Program.[31]

Writing to Beagle in March of 1967, Alfred Giardino, Board of Education Vice President, noted concrete reasons why the district was skeptical about the value of the MES Program. It was, he pointed, out very expensive, costing almost $500 more per child than was expended on children in the city's other elementary schools. MES schools, he went on to say, had not brought with them much in the way of curricular innovation, and were overstaffed.[32] Sidney Schwager, Chair of the UFT's More Effective Schools Committee, challenged Giardino concerns about costs. In an April letter to Giardino, Schwager accepted his figures concerning the cost of the MES Program. Yet, he felt that if this initiative helped children, it was "money well spent."[33] Not everyone on the board, however, agreed with Giardino's negative assessment. Aaron Brown stated that he differed with Giardino. As he saw it, achievement data was not the only measure of a program's effectiveness. There were important "intangibles" that affect children's attitudes and interests that are not as easily assessed but are of great importance. Such evidence, which was often found in the opinions of parents whose children were enrolled in MES schools pointed to a successful program.[34]

In April, Schwager attempted to enlist Mayor John Lindsay's support for the program. Writing to Lindsay, he noted that MES schools are the only schools in "disadvantaged areas of the city that were succeeding in educating the children. The program is receiving nation wide attention while New York City, where it originated, refuses to support or expand it." The program, he concluded was not perfect, but "New York city's ghetto children have never tasted the opportunity for success that a More Effective School offers."[35] Lindsay, as it turns out, was not all that much of a supporter. In June, a number of proponents of the MES Program at P.S. 83 in Manhattan wrote Lindsay and other city officials to complain about possible cutbacks in the initiative. Responding to their letter, he noted that it was the case that there were significant deficits in the reading and arithmetic skills of the city's elementary school students. Yet, he was concerned about the cost of the MES Program, which he noted in some cases was twice that of regular school programs, and other similar experimental initiatives.[36] The president of the school's parents association and its UFT chair wrote back to Lindsay indicting that they did not understand how he could voice concern for the education of the city's poor and minority children while not supporting More Effective Schools.[37]

Where the MES Program stood by mid year was uncertain. The school district's Executive Deputy Superintendent, Nathan Brown, wrote to

Shanker in May to assure him that there was no plan to eliminate any of the city's More Effective Schools.[38] In the same month, Schwager responded to a request to from Vangilee Hall, the President of the PS 113 Parents Association, for her school to join the MES Program. She noted that the board of education did not want to expand the program and that those who supported it and the union had to develop a joint strategy to convince the board otherwise.[39] Two months later, Charles Cogan, the President of the American Federation of Teachers, was reporting the comments of members of the board of education's MES Advisory Committee to the effect that Superintendent Donovan's cutbacks "pointed to an ultimate death of the project."[40] And in the same month, the New York City Council passed a resolution calling on the board of education to "desist" from ending the MES Program for at least six years.[41]

III.

In July of 1967, Board of Education President Giardino announced cutbacks in the MES Program including the elimination of an audiovisual teacher and a health teacher from each school and the transfer of the program's director to other duties within the central administration. In making these reductions, he had gone against the recommendations of the Joint Planning Committee that had initially established the MES Program, but was justified, he felt, by the concerns raised by the Center for Urban Education in their 1966 evaluation.[42] Shanker was clearly upset and sent telegrams to members of the board of education claiming that these reductions had violated an understanding that he had with the board and the superintendent to maintain the program for the next three years. The board, however, denied that any such arrangement had been made.[43] A reduction in the More Effective Schools program was only one issue of contention between the UFT and the board. The union was also concerned about their inability to agree on a wage contract for the coming school year and the unwillingness of the board to approve a policy that allowed teachers to remove disruptive students from their classrooms. Unable to settle these issues, the UFT called a strike in September. Lasting two weeks, the walkout garnered the teachers a wage increase but did not lead to an expansion of the MES Program.[44]

The actual role that the MES Program played in the strike is difficult to discern. MES proponents, not surprisingly, emphasized the importance of the program in precipitating the strike. Other argued that the key issues were teacher working conditions, and support of the MES Program was seen as something of a subterfuge to mask the union's efforts to improve the working conditions and salaries of teachers behind a concern for the educational achievement of minority students.[45] At the same time, however, the New York City Congress of Racial Equality (CORE) chapter opposed the strike, questioned the commitment of the UFT to racial integration, saw the MES Program as an obstacle to Black control of the schools, and preferred their own Black-led program to improve instruction.[46]

The same month that the strike occurred, the Center for Urban Education conducted another evaluation of the MES Program. At the outset, the head of the evaluation team, David Fox, noted that this study fell into a group of what he called "short term evaluations" that are undertaken in the early days of a program. As a consequence, the assessment did not provide a conclusive picture of the outcomes but rather was suggestive of what might be its impact. This report, as it turned out, was less favorable than earlier evaluations.[47] The comments of administrators, teachers, and parents as well as the observations of evaluators noted that most MES schools exhibited a positive climate "characterized by enthusiasm, interest, and hope, and a belief among all levels of staff that they were in a setting in which they could function." At the same time, however, the evaluators commented that MES schools had made "no significant difference" in the academic achievement of participating students. Students enrolled in MES schools did not achieve any differently than students enrolled in other special programs or in the schools that were used as controls in this study. The performance of students in reading and mathematics in some MES schools actually deteriorated over time. Where there was any positive change, it did not maintain itself more than a year or two. Concluding on a less than positive note, Fox noted that it was not likely that the program in its present form would enhance student achievement nor reduce what was seen as "the pattern of increasing retardation" as children advance through the grades in MES schools.[48]

At the end of his report, Fox offered an explanation for this situation, particularly for the disparity between the perceptions of the program and its impact on student achievement. The changes that had taken place in the MES schools, including the reduced class size and the use of cluster teachers, were largely administrative and organizational. These were changes that would no doubt lead teachers and outside observers to have a favorable opinion of MES schools. By themselves, however, these organizational and administration changes would not have an impact on the academic performance of students. Improving student achievement would require a transformation in instructional practices that did not seem to occur. In this regard, Fox pointed out that despite the reduction in class size, MES teachers often taught the same way as they would have in larger, more traditional classrooms. As Fox and his team saw it, the key problem was the inexperience of many MES teachers and their lack of training for working in the program. As a consequence, they were not able to take advantage of the program's organizational features to change instructional practices.

Not surprisingly, the UFT challenged the Center's evaluation. Writing the following year in the Center's own journal, *Urban Education*, Sidney Schwager, the Chair of the union's Committee on More Effective Schools, took as his starting point Fox's comment about the tentativeness of the evaluation. Although Fox's report was in effect an interim evaluation and not the final word on the MES Program, Schwager was afraid that the board of education would embrace its findings and use them as justification for phasing out the program. As he saw it, the aim of the board was to "destroy" the MES Program. Much of Schwager's criticisms were methodological in nature and focused on a host of measurement issues. He quarreled, for example, with the Center's judgment that MES students were not progressing adequately in either arithmetic and reading. It was, he argued, inappropriate to compare these children, as the Center had, with national norms that assumed that children would make one year's gain in achievement in one year's time. The normal growth rate for New York's urban children in the city's array of special programs was less. When that difference was taken into account, Schwager noted, it was the case that MES students were making significant progress.

Much of the conclusions that the Fox study made concerning the lack of instructional innovation was based, Schwager noted, on classroom observations. Yet he pointed out that the instrument used by those who observed MES classrooms had no standardized guidelines. Observers were left on the own to decide what was innovative and what was traditional. Beyond that, Schwager pointed out, Fox himself had noted that there was clear evidence in the report of changes in instructional practices in the MES Program. Schwager did not understand how any objective assessment could miss them. Schwager concluded his critique by acknowledging that the MES Program was not "perfect." There were numerous suggestions for improvement suggested in the Fox study that were worth considering. He did, however, wonder why the board of education had never undertaken such an extensive assessment earlier in the life of the program. Finally, Schwager noted that it was unreasonable to expect the MES Program in the short time that it had been in existence to solve all the problems facing urban schools. The most appropriate course of action, he argued, was not to end the program but for the board of education itself to conduct a more thorough evaluation during the next two years.[49]

For the remainder of 1967 and into the next year, the union continued to challenge whoever questioned the efficacy of the MES Program.

Speaking at the annul meeting of the American Federation of Teachers in August, Schwager questioned the accuracy and completeness of the Center for Urban Education's evaluation and noted that it had to be seen in light of the fact that the school district did not support the MES program and was seeking to get rid of it. He went on to criticize Superintendent Donovan for releasing part of the findings of the report to the press before it was actually completed, and he wondered if doing so had anything to do with the fact that the board was then in negotiations with district teachers over a new union contract.[50] In October, Schwager responded to an editorial on a local New York City radio station in which he criticized the media for "uncritically" reporting the "negative findings" of the Center's evaluation and that, he reminded his listeners, "experts in research have found serious fallacies in the methods used and conclusions stated." He went on in his response to call on the board of education to "cease its continuing attempt to destroy the More Effective Schools" and instead to undertake an effort to improve the actual weaknesses of the program.[51] The next month he wrote the City Club about their upcoming assessment of the first two years of the Lindsay administration to remind them of the ongoing effort of the board of education to undermine the MES Program and to point out that the Mayor's lack of support for the initiative was a "clear indictment" of his administration.[52]

In November, Superintendent Donovan submitted his budgetary request for the following year. In his letter of transmittal to the board, he made note of the need for the district to focus its expenditures on two problems, effective instruction and community participation. He went on to say that budgetary requests have been limited to those expenditures that directly affected the instructional program, specifically class size, teacher-student ratios, professional development, guidance, and curricular materials.[53] Supporters of More Effective Schools were clearly bothered by this action. The City-Wide Parents MES Association lauded the superintendent for the focus of the proposed budget on instruction. The goals that he spelled out were, they maintained, precisely the ones that the More Effective Schools Program sought to achieve. They were, however, upset that the budget did not include any funds to restore the cuts that had been made in the MES Program earlier in the year.[54]

IV.

Related to the conflicts between the board of education over the effectiveness and cost of More Effective Schools was another issue, that of race. In developing the MES Program, the Joint Planning Committee made the assumption that the aim of more effective education required that children of different ethnic and racial groups are educated together. They pledged that a key criterion for selecting schools for participation in the MES Program was their degree of racial integration. They in fact went further to claim that any educational effort to address the problems facing the city required "total integration in our schools."[55] In practice, however, integration did not occur. MES schools were located for the most part in the city's Black and Puerto Rican neighborhoods with largely minority group enrollments. Only four of the twenty-one participating schools had White student populations of over 25 percent, and only one the schools used bussing to promote integration.[56]

A good portion of the conflict over the program revolved around the issue of whether a separate, remedial program constituted the best way to enhance the achievement of minority group students. Writing in the *New Republic in* July of 1967, the syndicated columnist Joseph Alsop noted the desirability of ending the de facto segregation of urban schools. The dilemma, however, was that demographic shifts in urban populations had left most large city school systems without sufficient White students to allow for much in the way of integration. He went on to say that forcing integration under these circumstances would drive the few remaining Whites from the city thereby exacerbating the problem.[57]

The only viable solution to the educational problems of the city, according to Alsop, was "inside the ghetto schools." Challenging the conclusions of the Coleman report in favor of integration and the claim that Blacks could only learn in integrated settings, he called for programs that would attempt to improve inner city, largely minority schools. In this regard, Alsop cited the More Effective Schools Program as an example of the kind of effort that had a realistic chance of improving city schools and enhancing the attainment of minority students. He went on to chastise those critics of the program who questioned its added costs as well as those who opposed it in favor of what was in his mind a futile hope for school integration.[58]

In a response to Alsop three months later, Robert Schwartz, Thomas Pettigrew, and Marshall Smith questioned his assessment of More Effective Schools. Citing the Center for Urban Education's evaluation of the program, they challenged Alsop's claim that quality education was possible in racially segregated schools. Like the Center's report, they noted that despite the positive atmosphere that existed in MES classrooms, the program did little to improve the academic performance of its students. Schwartz and his colleagues were not opposed to compensatory education programs like MES. As they saw it, such programs represented a first effort, but the ultimate goal must be integrated schools. They did, however, bring with them, they argued, liabilities. They were very expensive, and it was not clear that city school systems could afford the added costs. A particular problem with the MES program, according to Schwartz, Pettigrew, and Smith, was that the necessity for smaller class sizes would require the building of new school facilities in inner city neighborhoods, which would have the effect of institutionalizing racially segregated schooling.[59]

In the end, it is difficult to say how successful More Effective Schools were in addressing the academic achievement of minority children. The five external evaluations of the MES Program that occurred in 1968 and 1969 pointed to mixed results. These assessments were more favorable than the 1967 Center for Urban Education evaluation that sparked much of the conflict between the board of education and the UFT over this initiative. Yet taken together, it was not always clear that MES schools had accomplished their principal mission of enhancing the academic achievement of the city's minority youth.[60]

The last of these studies, undertaken by the Psychological Corporation in 1969, sounded strikingly similar to the earlier and critical assessment by the Center for Urban Education. The evaluators noted that the MES Program was successful in reducing class size, in expanding the use of heterogeneous grouping, in increasing the size of school staff, and in increasing the availability of educational materials in the schools. They also lauded the program for instilling students with positive attitudes toward school and learning and with enhancing self-esteem. At the same time, however, the study pointed to the failure of MES schools to enhance the achievement of participating students in reading and mathematics.[61]

Despite these findings and earlier evaluations, other chapters of the American Federation of Teachers expressed interest in developing MES schools. Detroit established its Neighborhood Education Center that introduced the practices of the MES Program into a complex that included an elementary, junior high school, and high school on the city's East side.[62] Union chapters in Ecorse and Flint, Michigan contacted the UFT for information about the program.[63] In the midst of the long 1968 teachers' strike that resulted in the decentralization of the New York City's schools, the American Federation of Teachers National Council for Effective Schools visited locals in Des Moines, South Bend, Cleveland, Minneapolis, and Milwaukee among others to promote the establishment of a More Effective Schools Program in those districts. And a number of school systems throughout the country, including those in Newark, New Jersey and Santa Barbara initiated plans with their union locals to introduce More Effective Schools.[64] New York City's More Effective School Program would last in one form or another until the mid-1970s. Yet, in the decentralized system that emerged after the strike, it never played the central role that it had in the early 1960s.

V.

In October of 1966, Louis Schwartz, the Principal of Whitelaw Reid Junior High School (JHS # 57) in the Bedford-Stuyvesant section of Brooklyn, wrote the UFT's Si Beagle about a new program being established in his school under the auspices of New York University that embodied the principles that Beagle had thought essential to educating children in ghetto schools. Known as the Clinic for Learning and funded by the Ford Foundation and the New York City Board of Education and operated by the university, the Clinic was a multifaceted effort to use the school to improve an impoverished inner-city community. The Clinic brought university faculty into the school to provide remedial instruction, established a demonstration school to improve the university's teacher education program, served as a site for a program to prepare able high school graduates for teaching careers, and undertook other activities to enhance the skills of parents and to improve the surrounding community.[65]

As its proponents saw it, the Clinic was a cooperative venture between the city and the university to provide a model for high quality urban education. It would allow for the testing of new ideas and practices that would eventually become part of the regular curriculum as well as for the development of teachers who possessed knowledge of the problems facing urban public schools. The Ford Foundation was willing to support the initial costs of this project with the understanding that the board of education would ultimately assume financial responsibility for funding and replicating those innovations that proved successful.[66] The Foundation awarded the board a grant of $350,000 to inaugurate the program with the city providing $100,000 over the course of the next two years to fund additional non-university positions required for the project.[67] This seemed to be the kind of program that Cohen had in mind when he called for university-school partnerships as an alternative to the MES Program.

The impetus for the program was an array of educational problems that New York University had identified in a 1966 survey of public schools in Bedford Stuyvesant. The report noted that the schools were largely Black and Puerto Rican in ethnic composition, enrolled students whose academic performance, particularly in reading, was low, exhibited a high drop out rate among its enrollees, and sent a far lower proportion of its graduates to college than did schools in White, suburban communities. According to the report, these schools were failed institutions. The report noted in conclusion that every school in Bedford Stuyvesant was a "Special Service School," a designation that pointed to a school's low performance and its need for additional personnel and instructional support.[68]

The report included an appendix that identified important characteristics of Whitelaw Reid Junior High School. It was located in an extremely poor section of the community with marginal housing and high rates of juvenile delinquency. There was a high percentage of working mothers among the parents of children attending the school, and parent participation in affairs of the school was low. Not surprisingly, student attendance and academic performance were low.[69]

A central problem facing the school and affecting its ability to educate the community's youth was its organization. As Clinic supporters saw it, the typical classroom arrangement in JHS # 57 placed a large number of children in one classroom under the supervision of a single teacher. Under such an arrangement, the student was "simply a face in the crowd," unable to receive the kind of personal attention that would combat the "social and psychological disintegration" that was viewed as being part of the life of ghetto children. In its place, each grade in the new Clinic arrangement would be divided into smaller clusters. The seventh grade, for example, was divided into six clusters comprised of about eighty-five students each and six teachers. The staff included teachers in mathematics, English, science, and social studies, an additional undesignated teaching position, and a NYU instructor who would serve as the Cluster Coordinator. The Coordinator assumed a number of responsibilities including assisting the teaching staff in their instructional activities and in curriculum development, administering the guidance and counseling activities, and directing the transition to individualized instruction within the cluster.[70] The school would also serve as a training site for a variety of individuals preparing to teach in urban schools including undergraduate students enrolled in the university's teacher education program and able disadvantaged high school graduates and transfer students who wished to prepare for teaching careers.[71] The overall plan, then, was to demonstrate that with a different kind of teacher education program, a failing school in the ghetto could be transformed into a successful school. Children enrolled in such a school would reject their negative self image that had served to undermine their belief in the value of education.[72]

Establishing clusters would in effect create a number of smaller schools within the larger Whitelaw Reid School. And each of these clusters would turn out somewhat differently reflecting the interests and abilities of the teachers who staffed them. One seventh grade cluster class, for example, pursued a curriculum comprising the major academic disciplines while another cluster implemented an integrated core curriculum.

The Clinic also appointed a number of individuals from the immediate area surrounding the school as community agents who could establish connections between the families of children attending the JHS # 57. These agents would attend school meetings, establish contacts between the school and families, develop communication between the school and other community agencies, and provide emergency aid to students, parents, and school staff.[73]

VI.

Initially, the Clinic seemed to get off to a good start. Writing to Superintendent Donovan in October of 1966, Clinic Director John Robertson noted how much he respected both the school's principal, Louis Schwartz, and the district superintendent, Abraham Tauschner. Both of these individuals, he felt, were "committed educators who respond to every problem as a challenge to be overcome."[74] Yet, it proved difficult to coordinate the work of the school system and the university. In November, Robertson noted that it was important for NYU to have control over the school's instructional program, particularly where Clinic staff were attempting to move in new directions. He pointed out that the initial agreement for the project gave the university "carte blanche" regarding the direction of the curriculum. What was necessary, he went on to say, was for the central administration to allow the school's principal to "relinquish authority to the Clinic" concerning the curriculum. Such a shift in control, Robertson believed, was necessary "so that we can all continue to move as partners toward making important changes in education through the experimentation in Junior High School 57."[75] The challenge, he went on to say, was to recognize the legal authority of the board of education for the education of students while at the same time according the Clinic with responsibility for experimentation and innovation.[76]

This supposed arrangement and the latitude that it gave the university however, soon became a source of tension. Writing to Superintendent Donovan in February of 1967, Robertson reminded him of their agreement. He noted that the Clinic must be "free wheeling" in its ability to implement ways in which to educate disadvantaged children and that the public school personnel and university faculty working in the school must learn "to live with each other." Robertson then went on to describe several new initiatives that the Clinic had recently introduced or was in the process of implementing as part of a community development effort, including dental and optometric examinations and services and providing a group from NYU's School of Social Work to work with the families of students enrolled in JHS #57. Robertson also made note of a new grant that the Clinic had received from the U.S. Office of Education to support the training of cluster Coordinators for other city schools.[77]

Robertson's letter precipitated something of an angry response from the school's principal. Writing to Donovan and Robertson a couple of days later, Schwartz commented that the agreement for the establishment of the Clinic did not give the university free reign over the school's instructional program. He also took some offense at Robertson's suggestion that those public school and university personnel working in the program had to learn to "live with each other." Such a statement to his way of thinking implied that there had been efforts to undermine innovation and experimentation at the Clinic, a charge that he denied. Schwartz also had doubts about the Federal grant for training Cluster Coordinators. He felt that before new efforts to expand the Clinic's program were introduced, it was important to ensure that its original agreed upon activities were operating smoothly.[78]

These disagreements between NYU and the school's administration were part of a larger conflict over the purpose and direction of the Clinic. Interviews with university and public school administrators conducted by the Center for Urban Education as part of a 1970 evaluation pointed to this struggle. In his interview with Center staff, the Clinic Director noted a divergence of views between the school's administration and himself.

Robertson's initial impression when he arrived at the school was that it was an "oppressive place" that functioned "in a state of educational collapse." The school was part of the urban problem, and he did not think that it had the ability to regenerate itself. NYU, he went on to say, was more willing to change than the school. The school's teachers were a particular problem for Robertson. He stated that they lacked knowledge in their subject fields, were in some cases racists, and did not represent good role models for the students.[79]

The Center's interview with the principal pointed to equal dissatisfaction on the part of the school. Schwartz claimed that the Clinic staff operated outside of existing school procedures. He felt that the Cluster Coordinators did not spend as much time at the school as did the teachers. He was also bothered by the fact that Robertson, in his opinion, only visited the school infrequently. Overall, he was critical of what he saw as lax administration and the absence of any explicit procedures for making decisions. The interview with the principal indicated that something of a personal antagonism was developing between Schwartz and the NYU staff.

Schwartz was upset that during the first year of the program, his name had been removed from the Clinic stationery. He also objected to the practice of serving cocoa to students because it required that his janitorial staff had to clean up the spillage.[80]

The Center' interview with the district superintendent also pointed to the school's dissatisfaction with NYU. He was upset that the university did not consult either the district administration or the local school board in making decisions about the Clinic's program. Nor, he went on to say, did the NYU staff meet regularly with teachers assigned to the Clinic. The district superintendent was also critical of instances which pointed, in his opinion, to the "shortchanging of the brightest children." They included the absence of enrichment activities for children in the seventh grade and the refusal of the Clinic administration to allow the school's reading specialist to hold an essay contest on Black history until overall achievement levels were improved. Finally, he noted the complaint of one of the school's Black teachers about the failure of the Clinic to employ any African Americans on the staff, an indication to his way of thinking of a "missionary attitude" on the part of NYU.[81]

The Center's evaluation team also visited Clinic classrooms. In their words, this was a "demoralizing experience." They noted such disciplinary problems on the part of students as the lack of attention, talking, and refusal to follow teachers' directions. They reported that teachers seemed to be of the mind that students did not want to learn. The team went on to report that teachers were teaching subjects for which they had no preparation. One of the teachers they talked to during their visit described her efforts as "trying to put a band-aid on a great gaping sore."[82]

Not only was the school district critical of the Clinic. So, too, was the Bedford-Stuyvesant community. At a March, 1967 meeting of the Clinic's Advisory Board, local residents charged NYU with neither understanding the community in which the Clinic was located nor cooperating with school officials. Members of the Afro-American Teachers Association attending the meeting also voiced complaints about the Clinic. They noted that vacant positions in the Clinic were not filled, that students exhibited poor discipline, that no Blacks had been hired for the program, that no new curriculum had been developed for the clinic, and that the university students who were completing fieldwork at the Clinic were not carefully selected. Those attending the meeting were of the opinion that ultimately the Clinic would be used to provide justification for the belief that Black children cannot learn. Parents, as it turned out, were suspicious of the Clinic. They were worried that the school's program would reflect badly on the abilities of their children. They were not satisfied with the school before the Clinic began. The Clinic raised their expectations only to then dash them.[83]

Particularly troubling to parents was the production of a film about the Clinic. Entitled *The Way It Is,* the motion picture was ostensibly designed to promote the efforts of the Clinic to improve the educational program at the school. Yet, the reaction was far different. In a letter to NYU's Education School Dean, Superintendent Donovan complained that the creators of the film "cared more for the shock to society than they did for the truthful portrayal of the facts of public schools in disadvantaged areas." He labeled the film as "another *Blackboard Jungle*" that emphasized the turbulence and disorder of urban classrooms. He went on to say that the film was "a grave injustice to Negro and Puerto Rican children by giving the impression that they are unruly, uncontrollable and unwilling to learn."[84] A number of parents were so upset about the film that they requested the school district to remove the Clinic from the school.[85]

Despite the conflicts between the university and the school district, the proponents of the Clinic had plans for a second year of operation as well as a possible expansion. They hoped to be able to replicate what they thought were the best features of the Clinic and toward that end had made plans for the 1967–1968 academic year to introduce some of the Clinic's features in JHS # 118 in Manhattan. Those plans, however, were not to materialize.[86] In February of 1968, the school's Advisory Board appointed a committee comprised of five community members to conduct an evaluation that turned out to be very critical. According to the evaluation committee, PS # 57 had been in dire straights prior to the arrival of the Clinic. Student reading performance was below grade level. The Parents Association was not functioning effectively. And there was little in the way of guidance and related services available to the students. The goal of the Clinic was to use university staff and students to improve the operation of the school.[87] The committee noted that Clinic was established by New York University without any knowledge of the Bedford Stuyvesant community.

The committee looked at the Clinic from a variety of angles. There were major discipline problems including little in the way of building security to prevent those outside from entering the school. Students were often in the halls and regularly tardy to class, which resulted in ongoing disruptions from children wandering into classrooms late. Teachers seemed to have little control over the students and made little effort to manage them. They appeared to be afraid of the children, and parents exhibited little in the way of respect for the teachers.[88] They also observed a Clinic reading class that was held in a converted storeroom. The room had no windows or ventilation. There was no bulletin board and no pictures hanging on the wall. They reported that the teacher in this reading class had no lesson plans or a clear sense of what she was going to do each day. The teacher herself commented to committee members that she was not teaching reading skills and that the materials that she had available were primarily designed to enhance the students' self concept. The committee noted that there was little interaction between the Parents Association and the school. Finally they noted that the despite the role of the principal in running the school, the Clinic appeared to run itself. There was little communication or interaction between teachers and the Clinic staff. The evaluation group noted that during the Clinic's existence, student academic performance did not improve. In some areas it actually decreased.[89]

An exchange of letters between the school's principal and Clinic administrators in the midst of this evaluation seemed to seal the program's fate. Writing in January to Superintendent Donovan, Principal Schwartz noted that pressure was being placed from a source that he did not identify to continue the Clinic after June. The contribution of the Clinic to the school's program was, in his opinion, questionable, and he requested to begin planning for the next academic year without its presence in the school.[90] A few days later, Lewis Clarke, the Clinic's Resident Director, wrote to Donovan with his own assessment of the situation at the Clinic. He challenged Schwartz's claim that behind the scenes pressure was being exerted to keep the Clinic open. The Clinic's staff and the school's Parent Association, he went on to say, had indicated a desire to continue the program. These efforts were made publicly. This was not, as Clarke's letter seemed to imply, a "sub rosa plot."

Clarke then went on to address Schwarz's charge that continuing the program served no useful purpose. He described a number of initiatives undertaken by the Clinic including small group instruction in mathematics and English for the majority of the program's students, the organization of staff into clusters to enhance cooperation among the staff, the introduction of changes in the Parent's Association, and assistance with discipline problems in the school. The biggest problem facing the Clinic, according to Clarke, was Schwartz's "pervasive negativism." The principal's judgments to Clarke's way of thinking were not based on the actual contributions of the Clinic but were "biased and unprofessional." He was especially troubled by the fact that Schwartz penned his letter in the midst of the evaluation being undertaken by the Advisory Board and had the potential to undermine the objectivity of those assessments. He concluded with the observation that an experimental program like the Clinic could not succeed without the backing of the principal. He believed that there existed a "limited working relationship" with the school's administration but that "the overall climate is 'unhealthy'."[91] Writing to Superintendent Donovan early in February, Robertson urged that that the school district take into account the evaluations of the Clinic then being undertaken before deciding on a course of action.[92] As we have seen, the evaluation was decidedly negative and called for the discontinuation of the Clinic, which occurred at the end of the 1968 academic year.

VII.

The conflicts over the More Effective Schools Program and the Clinic for Learning represented two of many battles during the 1960s that divided Black and Puerto Rican New Yorkers from the city's teachers, the UFT, and the board of education. At the heart of this division was the failure of Black and Puerto Rican New Yorkers to secure the kind of racially integrated schooling that they believed would improve the educational prospects of their children. Early in the decade Black and Puerto Rican parents had charged the board of education with betraying their interests when it had shifted the location of a proposed intermediate school (I.S. 201) from a site near the East River just across from Queens to what became its actual location in the middle of Harlem. This change of location had blocked what the board had claimed was their intent of creating a model integrated school comprising Black and Puerto Rican students from Harlem and White children from Queens. Further infuriating these minority parents and activists was the decision of the board when the school opened to appoint a White principal.[93]

The response of an I.S. 201 parent-community committee to a set of 1966 proposals that the board of education made for improving education in disadvantaged schools throughout the city pointed to the increasing chasm between these two groups. The proposals noted that the longstanding efforts of the city's schools to provide remedial programs for low-achieving minority children had brought little in the way of improvement. It went on to suggest that the criticism that some Black and Puerto Rican parents had voiced about the competence and dedication of teachers was unjustified. As its authors saw it, the achievement problems of ghetto schools were not simply the result of poor teaching and the absence of adequate educational resources. Also critical were the economic distress and deteriorating physical conditions of the communities in which many minority children lived and the seeming inability of their parents to be effective advocates for their children.[94]

Two features of these recommendations raised the committee's ire. They were upset by what they saw as its implicit message that attributed educational failure to the "inherent inadequacies in Black and Puerto Rican communities." Such an explanation, they claimed, ignored the responsibility of the schools themselves for their failure to educate minority children. They were also upset about one of the report's recommendation for the establishment of an advisory board to develop recommendations for the improvement of I.S. 201 and its feeder schools. Such a panel with only advisory authority would lack real power to change the conditions of the schools but would deflect the responsibility that the district and the board of education bore for the academic failure of Black and Puerto Rican children. It was a report, according to the committee, that "takes its place at the top of a large pile of similar proposals as this season's document of deceit and duplicity from the 'Board of Genocide' of the City of New York."[95]

By the mid-1960s, Black and Puerto Rican parents and community groups came to lose faith in the prospect of racial integration and to embrace in its place efforts to improve conditions for teaching and learning within what were ostensibly segregated schools. Improving school conditions was, however, only part of their goal. In early 1968, the Brooklyn and Bronx chapters of CORE along with the Bronx Community Self Improvement Association drafted a plan for the creation of the Institute of Learning to provide an alternative to existing board of education policies for dealing with students who were viewed as being disruptive. Such students were typically either expelled from regular schools or transferred to special programs for the maladjusted or emotionally disturbed.

The Institute's proponents did not believe that there was really anything wrong with these children. "The so-called "disruptive child" is usually," they argued, "a high spirited, non-conformist, who is not willing to accept the mediocre education available in most ghetto schools."[96] The supporters of the Institute did not think that the kind of remedial programs that the board of education offered to such students was appropriate. What was called for, they maintained, was an enriched program that provided them with a challenging curriculum related to their day-to-day lives. They proposed, for example, sewing courses in which students would be able to design and make clothes that they could wear and science classes in which students would explore such "real life problems" as how to "scientifically eliminate rats from their community."[97]

The board was initially receptive to the idea. What troubled them, and eventually led them to withdraw support was the large role that the plan gave to parents and community groups in the governance of the Institute. Writing to Robert Carson, the community relations director of Brooklyn CORE and a key supporter of the Institute, Superintendent Donovan noted that this proposal was to be a cooperative effort with the community but that the school "will remain under the supervision of the Board of Education and will be staffed by the Board of Education with the assistance and guidance of your organizations."[98] Donovan's letter elicited an angry telegram from Solomon Herbert, an official of the Bronx CORE chapter noting that his letter to Carson did not reflect the "content and tone of our proposal" and demanding an immediate meeting with Donovan.[99] Donovan responded by refusing to meet and indicated that his letter to Carson clearly indicated "the exact nature of our relationship in this project."[100] Two days later, the Brooklyn and Bronx chapters of CORE telegraphed Donovan indicating that they were abandoning the project because they could "no longer maintain or continue any kind of relationships with people who have no integrity, sincerity or professionalism."[101]

CORE's quarrel with the board of education over the Institute of Learning was a dispute over who should control the education of Black children in New York City. Was such authority the purview of professional educators or did it lie with African Americans themselves? The answer for increasing numbers of Black New Yorkers was the latter. Carson himself addressed this issue in his autobiography in describing the efforts of parents of children enrolled in Junior High School 258 to remove a principal who they viewed as incompetent. "The Community, not Donovan," Carson argued, "should have that responsibility. Anyone working in Our Community with our children should be accountable to that community and, as far as Brooklyn CORE was concerned, the request by the parent of that school gave us the authority to do whatever necessary to bring about the relocation of that principal."[102] As Carson saw it, increasing numbers of Blacks were dissatisfied with the quality of education that their children were receiving in the city's school, which was leading them to talk about the need for community control.[103] More broadly, it was Carson's view, as he put it, "that the community of whatever race or mixture of races should determine how its children should be best educated."[104]

The conflicts surrounding More Effective Schools and the Clinic for Learning that I explored in this chapter were two phases of a larger twentieth century movement on the part of African Americans, marked most pointedly by the 1954 *Brown vs. Board of Education* decision, to challenge school segregation and to promote equal educational opportunity. The impact of the decision on school desegregation has at best been mixed and its larger effect on equality of opportunity disputed. Yet, for many but certainly not for all Americans, the decision was one of those singular events that marked the possibility of fundamental change on the American landscape.[105] In commenting on the decision the novelist Ralph Ellison noted "what a wonderful world of possibilities are unfolded for the children.[106] And in his history of the Brown decision, *Simple Justice*, Richard Kluger viewed the ruling as "nothing short of a reconsecration of American ideals."[107] In terms of the argument being advanced in this volume, the decision seemed to signal the possibility that with the movement toward racial integration, Americans might come together in a spirit of community that would bridge the nation's longstanding racial divide.

Yet, as we have seen in this chapter, that sense of common purpose and collective belonging never emerged. The next ten years was marked, particularly in the nation's largest cities, by the kind heightened level of racial conflict that we saw in the struggle surrounding More Effective Schools and the Clinic for Learning. Such discord would pose a challenge to the aspirations of Blacks for full citizenship and participation in an integrated society. It would bring to the fore at the behest of proponents of Black Power and Black Nationalism a very different vision of African American identity rooted in ideals of racial separation and autonomous development.[108] In our next chapter, we look at one instance of how this challenge developed and played itself out in Detroit, Michigan during the mid-1960s.

Chapter 3

Community, Race, and Curriculum in Detroit: The Northern High School Walkout

On the morning of Thursday, April 7, 1966, some 2,300 students at Detroit's all Black Northern High School responded to Superintendent Samuel Brownell's decision to close the school in anticipation of a student protest by walking out in mass and joining a group of parents who had congregated on the street in front of the school. For the next two hours, the students and parents marched around the school carrying picket signs that decried the education offered at Northern and shouted for the removal of the principal, Arthur Carty. The marchers then made their way to nearby St. Joseph's Episcopal Church where about a thousand of them attended a rally in which they listened to a recent Northern graduate tell how his high school education did not prepare him for the academic rigors of the University of Michigan and heard Superintendent Brownell's assurances that the problems at Northern would be addressed.[1] What followed was a three-week boycott during which time the student leaders of the walkout would press their demands for a host of changes at the school, including the removal of the principal.

Over the years a number of accounts of the Northern walkout have appeared in print. Within a year of the boycott, two of the key participants, Rev. David Gracie, the rector of St. Joseph's Church, and Karl Gregory, a Wayne State University faculty member and 1947 graduate of Northern, offered separate recollections of the protest and its aftermath.[2] At about the same time, the National Education Association's Commission on Professional Rights and Responsibilities issued a report of its investigation of conditions in the city's schools, which devoted space to a discussion of the boycott.[3] In his history of the Detroit's 1967 race riot, written some two decades after its outbreak, Sydney Fine looked at the Northern High School walkout as one of the precursors of the civil disruptions that broke out in the city in July of that year.[4] And most recently, Jeffrey Mirel in his history of the Detroit Public Schools during the twentieth century saw the walkout as evidence of the breakdown of a common educational vision that had emerged among White and Black Detroiters in the previous decade.[5]

Despite the differences in these interpretations, they all viewed the walkout as an indication of Black dissatisfaction with the education that their children were receiving in the city's schools and evidence of the growing divisions between Blacks and the largely White corps of teachers and administrators who ran Detroit's schools. Racial division, however, was only part of the story of the Northern walkout. At the same time, however, the boycott and what followed was the kind of crisis that, as the sociologist Robert Booth Fowler sees it, spurs forward among those involved feelings of common purpose, self- identity, and shared understanding.[6] In chapter one, I made mention of how race can provide the soil for cultivating such sentiments. And as we shall see in this chapter, those dispositions would serve a unifying role in helping to construct and shape a sense of community among the city's Black population.

II.

The precipitating event behind the Northern boycott was the refusal of the head of the English Department, Thomas Scott, with the support of the principal, some two weeks before the walkout, to allow publication in the student newspaper of an editorial critical of the school and the education it offered Detroit's Black youth. Written by a senior honors student, Charles Colding, and entitled "Educational Camouflage," the article pointed to the failure of urban schools like Northern to provide a quality education. Colding decried such practices as social promotion, which he claimed was responsible for the low-achievement of Black students at Northern. He argued that the underachievement of Northern students when compared with the performance of students in such largely White Detroit schools as Redford High was unacceptable. Finally, he blamed conditions at Northern, which he asserted were not accidental, on segregation and on what he claimed was the widespread belief among Detroit educators that "Negroes aren't as capable of learning as Whites."

On the Monday following the walkout, Superintendent Brownell met with the student leaders of the boycott to discuss their demands. Carty, they argued, should be removed as principal and Assistant Principal George Donaldson should not replace him.

The school administration, they felt, was paternalistic and authoritarian in their relations with students and the community, had ignored the inadequacies in Northern's curriculum and physical plant, and was responsible for its educational failings. Students further demanded concrete information that would allow them to compare the academic standards existing at Northern with those of other Detroit schools. They wanted the board of education to furnish them with a plan detailing how the problems at the school could be corrected. They also called for the creation of an elected student-faculty council to examine school problems and to make recommendations for their solution. The students asked for the appointment of a school-community agent to work with the parents of Northern students. They demanded the removal of the police officer who patrolled Northern because of his perceived harsh treatment of students. And finally, the students wanted guarantees against reprisals for those participating in the walkout.[7]

During the next few days, Superintendent Brownell tried to end the walkout by temporarily reassigning Carty to the school district's central office to assist the Northern investigation and placed two district administrators in charge of the school.[8] The students returned to school at the beginning of the following week, but a series of meetings between the board of education, the student protesters, and Northern's administration that were held to settle unresolved issues surrounding the walkout produced additional tensions. In the midst of these meetings, Brownell appeared to renege on his promise to the students and announced that Carty would be returning to Northern. This produced a second walkout that involved all but about two hundred of the school's students.[9]

This protest lasted about a week during which time the vast majority of student boycotters attended a newly established Freedom School that Karl Gregory had established at St. Joseph's Church to compensate for what he saw as the inadequacies of Northern's curriculum.[10] A series of meetings involving the protest leaders, prominent community members, and school authorities during the second walkout finally produced an agreement on April 25. Brownell reassigned Carty to the district's central office, where he would work on resolving the school's problems. He would, however, continue to keep the title of principal although George Donaldson would become acting principal for the remainder of the year. In addition, a citizens committee would be appointed to undertake an investigation of conditions at Northern. The next day, the students returned to school, and the boycott came to an end.[11]

III.

The refusal of Northern's English Department Head to allow the publication of Colding's editorial precipitated the walkout. Censorship, however, was only part of the story behind the protest. The Northern boycott, as it turned out, was not an isolated event in this system of 294 schools enrolling approximately 295,000 students.[12] A number of Black students from Eastern High School joined Northern's second walkout on April 22, and toward the end of the month about seventy students from Southeastern High School held their own one-day boycott voicing similar complaints to those made by the Northern students.[13] Nor were these the first such boycotts. In September of 1947, Black parents took their children out of Higgenbotham Elementary School for a period of almost three weeks to protest what they saw as the board of education's attempt to maintain segregation by transferring Black children out of racially mixed Post Intermediate School to Higgenbotham. Twelve years later, in November of 1959, parents organized a one day student boycott at Pattengil Elementary School in response to the decision of the school administration to relieve overcrowding at the largely Black school by transferring seventy-four students to another predominantly Black school although two schools with majority White student bodies were closer.[14]

Student walkouts were a recurrent form of dissent that the city's Black population employed to protest the existence of racially segregated schools, the decidedly inferior education that they believed such schools provided, and the resulting pattern of persistent low achievement that affected their children. At the root of these conditions were the demographic shifts and economic changes that occurred in Detroit and other large American cities in the two decades following World War II. This transformation had spurred forward the mutually reinforcing factors of residential and employment discrimination, suburbanization, deindustrialization, and a racialized politics that taken together led to a significant decrease in the city's White population coupled with a pattern of plant relocations, downsizing, job loss, and capital flight. The result was the destruction of the city's manufacturing infrastructure, the erosion of its tax base, persistent poverty, and economic dependency.[15] By 1960, Detroit had a population of about 1.6 million, down from its high point of almost 1.9 million ten years earlier. Between 1950 and 1960 the Black population had increased from 300,000 or about 16 percent of city residents to almost 500,000 or 29 percent. During the same period, the city's White population declined by about 23 percent.[16]

The Northern protest was one instance of this longstanding conflict between Black Detroiters and the largely White school authorities over the education that the city offered its African American children. The most detailed statements of Black dissatisfaction with Northern were the essays that students attending the Freedom School wrote describing the problems at the school. Several essays noted that teachers at Northern did not care about and in fact disliked African Americans. As one student put it, Northern teachers "look and teach down to Negro students." In the words of another student, teachers at the school believed that "Black boys and girls don't want to learn, so therefore they don't put much in their jobs." The principal fared no better in their assessment. Carty, they stated, did not know what was going on at Northern. Nor did he care. They felt that he ran the school like a "dictatorship" and that he was "not for the Negro."

The students claimed that there was an absence of books and equipment at the school and that the facilities were inadequate. Student essays noted the availability of six working sewing machines in a sewing class enrolling thirty students, broken typewriters in the typing room, and a swimming pool with water that was unbearably cold. The essays point to the inadequacy of the curriculum. A student reported that there was little difference in what was taught to the advanced students in the accelerated English class and in the general track English class. One essay made the point that it was assumed that African American students were "willing to accept anything, the leftovers, and this is what we are given at Northern." In summary, as one student put it, "everything at Northern is either inferior, incompetent, or an injustice."[17]

There were teachers at Northern who lent credence to the students' claims. Paul Richards, a social studies teacher, pointed out that the administration had not been attentive to the needs of students and teachers. Further, he felt that a negative attitude to the effect that "Negroes can't learn" pervaded the school. English teacher Claudia Cullen alleged that Carty managed the school in an arbitrary manner. She thought that the walkout demonstrated that Northern students were concerned about getting a quality education. Another English teacher, Patricia Wieder, believed that good teachers at Northern were quick to seek a transfer to another school.[18] Similarly, in a letter to the *Detroit Free Press*, a teacher at a nearby high school anonymously charged that "unfit teachers" and the "constant humiliation" of students were undermining Northern's ability to provide an adequate education.[19]

Parents of Northern students and others who lived in the surrounding area were also critical of the school and its leadership. The citizens committee appointed by Superintendent Brownell to investigate the boycott held two public hearings a month following the end of the walkout, one attracting 25 participants, the other 75. Those who spoke accused

Carty of prejudice toward Black students. They were equally critical of Northern's teachers, who they claimed did not hold out high expectations for African Americans. Community residents charged that the school's teachers employed outdated teaching methods and did not seem concerned about the well being of their students.[20]

Black community leaders also cast doubts about the quality of the education offered at Northern. The Ad Hoc Committee of Clergy, chaired by William Ardrey, the minister of St. Paul AME Zion Church, issued a statement toward the end of the boycott supporting the demands of the Northern students. As these ministers saw it, Detroit's inner city high schools, including Northern, did not provide students with the skills that they needed for work or college. The cause of the problem was not, they argued, the "capacity of students to learn" but rather the "end product of an inferior educational program that begins in the elementary grades." What troubled them was that little was being done to correct the situation.[21]

On the other side stood Arthur Carty, Northern's principal, who saw things quite differently. He was critical of Colding's editorial and had told the school's English Department Chair that Colding "doesn't have his facts." As Carty saw it, the school offered its students an appropriate educational program. "We have all the courses they need here, if the kids want to take them." Whatever problems existed at Northern, he argued, had nothing to do with the school but were the result of the lack of student motivation coupled with poor parental involvement. "I have," he noted, "preached in churches and gone to block club after block club trying to persuade parents their children should work up to their full potential." Yet, very few parents took the time, he pointed out, to visit the school or attend school meetings.[22]

A number of Northern teachers seemed to agree with Carty about the situation at the school. James Samples drew up a petition supporting the principal that praised him for offering the student body "the best possible educational programs" and for working hard to enhance relations among the students, faculty, and community. Another Northern teacher noted that with Carty as principal, "things have improved here enormously." And history teacher Millicent Wills stated that not only teachers but also students and their parents had to accept responsibility for the situation at Northern. The blame, she went on to say, should not be placed on Carty who, in her words, "deserves better treatment."[23]

Charles Lewis, principal of Detroit's Central High School, shared Carty's views about Black students. His school, like Northern, was virtually all Black and included, according to Lewis, a large number of students who "will be truant, will be indifferent to learning and to failure, will often rebel against the meagerness of their lives and will strike back." Such students, according to Lewis, were intellectually normal but exhibited academic deficiencies. A major problem facing these students, Lewis argued, was that the existing high school curriculum, which was designed for those with aspirations for college and for middle class living, did not appeal to them. What was called for, he noted, were courses that recognized that "high school is terminal education for many people and each graduate should develop saleable skills." In this vein, he cited a new English course that had recently been created for the city's high schools and was being directed at students of average ability who were thought to have little in the way of academic interests and were seen as potential dropouts. Students taking the course read popular paperback books of the day, did their writing in a personal journal, and developed their speaking skills in discussions of issues that were thought to be of interest to adolescents.[24]

IV.

The Northern High School walkout was one phase in a larger pattern of conflict over education in Detroit that involved its White and Black citizens. One issue that was important to Blacks was curriculum differentiation. They saw this practice as a major reason why the city's largely Black schools offered a different and what they believed to be a distinctly inferior course of study to that provided in schools in Detroit's White neighborhoods. In 1958, Superintendent Brownell established the Citizens Advisory Committee on School Needs (CAC) as a means of ensuring more citizen participation in school policy making.[25] In its report, the CAC endorsed the longstanding practice within the city's schools of curriculum differentiation that channeled students on the basis of their supposed abilities to an array of courses and programs, some preparatory, others terminal. Assigning students to different curricular programs that were thought to match their abilities, the CAC believed, offered a way for administrators to reconcile the competing demands facing the city's schools. It would signal their intent to provide all of the city's children with equal educational opportunity through a set of common goals. At the same time, however, the availability of diverse classes and curricular programs indicated their commitment to meeting the particular needs of individual students.

According to the consultants who worked with CAC's Curriculum Sub-Committee, "wide differences in content and method may and probably should be evident between different classes of pupils of the same grade." This was the case, they went on to say, because "socio-economic and other factors operate in different regions of the city to make a uniform course of study unsatisfactory."[26] For the high schools, this view legitimated the sub-committee's continued support for a program comprising four courses: college preparatory, vocational, business, and general.[27] And for low-achieving children, the principle of differentiation supported their recommendation for the establishment of separate remedial courses in reading and arithmetic.[28]

Two years later in 1960, the Citizens Advisory Committee on Equal Educational Opportunity (EEO), which was appointed by the board of education to investigate the existence of discriminatory practices in the city's schools, also expressed support for curriculum differentiation.[29] Taking their cues from James Bryant Conant, a former President of Harvard University and nationally known expert on high school reform, the committee defined opportunity in meritocratic terms. Doing so, they made clear, did not mean that all children should receive an "equal amount of schooling." Rather, it meant providing the child with "the kind and amount of schooling . . . that his capacities warrant."[30]

With this as its starting point, the committee easily affirmed the principle of curriculum differentiation. "It may seem something of a paradox," they noted, "that equality can be achieved only by providing unequal education; but this paradox becomes clear when we realize that the students start school unequal in ability and experience, and that to offer them merely the same education in each case would be simply to continue the original inequality."[31] Consequently, they recommended the continuation and extension of the policy of ability grouping, the development of vocational and work experience programs along with efforts to ensure that Black students had access to such programs, and the establishment of remedial classes and programs to address the reading and arithmetic deficits of students.[32]

Despite the support that both committees gave to curriculum differentiation, a survey undertaken by the CAC in 1958 suggests that Detroiters were divided on the issue. The policy had its proponents. L. Glen Shields, an employee of the City's Department of Building Engineers, thought that it would be "foolish" to place children of very different abilities in the same class with the same assignments. "To force the dullard to compete with the potentiality of the brilliant is nonsense. To hold the fast learner to the pace of the slow is stultifying, discouraging, and a waste of money, and time." He felt that students who lacked the ability to do well in the basic subjects should be sent to special schools to learn a trade.[33]

Curriculum differentiation also had its opponents. Detroit's Urban League supported heterogeneous grouping. The League argued that the public needed to recognize the existence of individual differences and to understand that such diversity is desirable in a democratic society. The

League's spokesperson then added that the chief concern of the organization is the Black student who heretofore had been poorly served by the schools. The schools, according to the League, should attempt to maximize the potential of all students and aim toward the "equalization of competitive skills." And the best way to achieve that end, the League believed, was through heterogeneous grouping.[34]

Most of the respondents to the questionnaire, however, took positions somewhere in-between. Both of the city's teacher organizations, the Detroit Federation of Teachers (DFT) and the Detroit Education Association (DEA) supported special classes or schools for disabled children. For other children, the DFT advocated grouping policies that took into account the particular needs of different schools, and the DEA favored policies that ensured that children placed in special classes and groups had the chance to rejoin the regular group.[35] The local chapter of the American Association of University Professors also favored homogenous grouping within the various school subjects, but they did not want homogeneous schools. They believed that each city school should enroll children with a diverse array of abilities.[36]

R.J. Browne, Jr., representing the Mormon Church in Detroit, argued that the city's schools should make efforts to identify gifted and retarded students but should not segregate them. E. Dennis Arwood, speaking for the Michigan Society of Professional Engineers, suggested that initially children should be grouped heterogeneously to determine their abilities. But once that identification was made, they should, he argued, be grouped homogeneously to "accelerate the gifted ones."[37]

Another recommendation that got a mixed reception was the EEO Committee's school transfer policy that would allow students to move to any city school that offered programs not available to them in their neighborhood schools.[38] This too was an important issue for African Americans since they saw it as a way to propel forward integration while at the same time compensating Black children for the inadequate programs offered in their neighborhood schools.

A March, 1962, *Detroit Free Press* survey of the reaction of city residents to the report of the committee pointed to a wide difference of popular opinion concerning this recommendation that cut across racial lines. R.H. Weaver, a Black dentist, indicated that he supported the committee recommendation to allow students throughout the city to transfer freely from school to school. He went on to say that existing residential segregation virtually ensured that neighborhood schools would be segregated. Robert Jackson, a Black junior high school teacher, made the same point with respect to teachers. He argued that Black teachers, heretofore, had faced discrimination in city schools. Jackson believed that more teachers, particularly White teachers, should be assigned to Black schools and that doing so would serve to help the children enrolled in these schools. Nathan Garrett, a Black accountant, indicated that he was aware of the problem of segregation. Yet, he was not sure that he wanted Black children to attend school outside their neighborhood. Black teachers, he thought, would do the best job with Black children.[39]

White opinion was also divided. W.B. Spaulding, a White, retired tool and die maker who lived in a neighborhood that had become all Black, argued that the public schools should be open to all children. Another White tool and die maker, Henry Lake, took the same position. But B.A. Carlen, a White industrial engineer, on the other hand, disagreed with the committee's recommendation that allowed students to transfer to schools outside their attendance area. It would, he thought, cause too many people to want to move from this school to that school. Mrs. Florence Hollander, a White homemaker, agreed with Carlen about the value of the neighborhood school. She was, however, not opposed to assigning Black teachers to largely White schools.[40]

It was also the case that the members of the EEO were not united in their views. In a June, 1960 meeting of the EEO Committee, one member, Dr. Horace Bradfield, a Black physician, challenged a committee progress report that claimed that the majority of Detroiters believed that the city provided equal educational opportunity to everyone. As he saw it, African American children in Detroit were subject to discrimination. "By virtue of being colored," Bradfield claimed, "every Negro child is denied equal educational opportunity."[41] He went on to say that many of the city's teachers were not adequately prepared for their positions, that they did not treat Black children and their parents with respect, and that they were not interested in teaching Black students.

As a remedy, he recommended allowing African American children to attend any city school in any neighborhood, appointing Blacks to top school district posts, and restricting administrators to a maximum of ten years in their positions. Another committee member, Hubert Eiges, a White teacher at Mackenzie High School, took issue with Bradfield's charges. He claimed that Black and White children were treated the same in the city's schools. The problem, as he saw it, was that Black children "do not seem to participate even though a definite effort has been made in their direction." Eiges believed that racism or discrimination was not the issue. "It's a problem," he argued, "of human relations and a lack of motivation."[42]

The issue of equality came to the fore again three months later in September when the committee heard from one of its consultants, the psychologist and nationally known supporter of school desegregation Kenneth

Clark. Clark observed early in the meeting that much of the discussion thus far had avoided talking about what was in his opinion the major form of educational inequality affecting Detroit, school segregation. The committee chair, Juvenile Court Judge Nathan Kaufman, responded to Clark by saying that the members did not know at this point whether segregation existed in the Detroit Public Schools but would as one of their first orders of business attempt to find out.[43] Kaufman's comments led another member, Mrs. Claude Moore, to note that there were schools in the city that were segregated by race as a result of housing discrimination. A Black committee member, Ramon Scruggs, followed by stating that segregation and inequality were facts of life in Detroit and had been for at least three decades. He went on to say that he thought that Judge Kaufman was being evasive in not recognizing this situation.[44]

V.

To understand these educational tensions in Detroit and how they would play themselves out in the aftermath of the Northern walkout, we need to recognize two longstanding features of race relationships within the city. One was the increasing pattern of discord between Detroit's Blacks and the city's White working classes that accompanied the growth of Detroit's African American population in the years following World War II and the accompanying demographic and economic changes that we described earlier in this chapter. Blacks, they feared, would take their jobs, lower their property values, and infest their neighborhoods with crime and vice. Limiting Black employment opportunities in the city's automobile industry and maintaining residential and school segregation were the bulwarks of what they saw as their defense strategy. Whites who employed these tactics routinely sought to dissuade supporters of equal employment opportunity, housing integration, and school desegregation by both admonition and intimidation. When those efforts did not work, and they often did not, some Whites, particularly those who felt most threatened by Blacks, turned to outright violence. At the same time, numerous White politicians in Detroit sought to advance their own careers by abetting this White resistance.[45]

This White dislike and fear of Blacks came through loud and clear during the Northern walkout. A letter that a Mrs. C.R.F. penned to School Board President Remus Robinson, who was the first Black elected to the Detroit Board of Education, during the boycott is indicative of this attitude on the part of many White Detroiters. She pointed out to

Robinson that Detroit's taxpayers were tired of the "constant picketing and complaining" by the school's students and their parents. She asked Robinson how many of those involved in the boycott were like her, workers and taxpayers. She went on to say that the boycotters were largely Southern migrants to Detroit who upon arriving went on welfare with the expectation that everything they needed would be provided for them. As she put it, "colored people as a rule don't care to work. Just sit around and collect welfare checks." She went on to claim that Detroit had been a "clean city until people from [the] South came." Detroit, in her words, was the "worst city in high crimes" with "colored prostitutes" all over and with "killings, robberies, assaults, and rapes on White women by colored men."[46] Several years earlier Robinson had received a similarly virulent letter, this time from a White female student at the University of Michigan, who was upset about his support for the integration of the school's Greek organizations. "Niggers and kikes in our sororities and fraternities," she complained, "indeed!" She went on to say that he should "forget his stupid integration ideas! We're for segregation and keep your punks in their place."[47]

Blacks were, it seems, well aware of the enmity that many White Detroiters held out for them. The remarks of an African American teacher at Northwestern High School in 1962 characterizing the attitudes of the White principals who ran the city's secondary schools make the point. These administrators, he argues, had little interest in educating Black children. Rather, their driving motive was to please their superiors at the board of education and in so doing to earn a promotion out of the city's Black schools:

> A "good principal" will do whatever his bosses downtown ask him to do. The more a principal cooperates the sooner he will accumulate enough "gold stars" next to his name to move out of the depressed Negro school and into one of the lily-White schools (which will naturally be closer to his home as all of Detroit's high school principals are White). This means that he will keep cheating them every day in nearly every way. This means that he is guilty of criminal negligence, for a day lost in a kid's education is a day lost forever.[48]

Another feature of these relationships was the divisions that existed among Black Detroiters. Facing a hostile and antagonistic White community, the city's Blacks often forgot their own differences to unite against the common threat.[49] Yet, we should not assume that Black communities were cohesive or unified on the inside. There were in fact longstanding divisions among Black Detroiters. During the nineteenth century Blacks were divided into elite, middling, working, and poor caste-like groupings on the basis of their place of origin, education, and wealth, which determined their residence within the city, club memberships, church affiliations, and interactions with Whites.[50] Labor's early struggle to organize Detroit's developing automobile industry during the first forty or so years of the twentieth century was another source of division among Blacks. On one side stood the city's more privileged Blacks, a number of prominent Black clergymen, and organizations such as the Urban League and the National Association for the Advancement of Colored People (NAACP). They welcomed the limited opportunities that such automobile executives as Henry Ford had made available to African Americans and consequently viewed unions with suspicion. On the other side were more radical Blacks who supported the efforts of the fledgling United Auto Workers to construct an interracial brand of unionism comprising not just skilled workers but the unskilled and semi-skilled as well.[51]

During the 1960s, education was a key focal point for such Black-Black divisions. There were those who supported the continuation of a differentiated curriculum if it ensured opportunity for Blacks. In May of 1967, a number of residents from the neighborhood surrounding Northeastern High School appeared before the board of education to protest the decision of an administrative committee to close the school's machine shop and substitute in its place courses in gardening, landscaping, and institutional housekeeping. What upset community residents was the fact that these proposed courses were terminal in nature and would not allow students who took them to continue their education after high school. They were also troubled by their belief that students completing these courses would have difficulty finding employment in an industrial city like Detroit. These Black Detroiters were not opposed to the existence of a vocational curriculum. They did, however, want a curriculum that would produce skilled and semi-skilled workers among the city's African American population.[52]

There were other Blacks who wanted a more academically oriented education for their children, which they claimed was available to the city's White children but denied to theirs. In October of 1966, the newly established Higgenbotham Elementary School Parent and Citizens Committee threatened to boycott the school in protest of what they saw as the inferior education offered their children. They were upset about the fact that the achievement test scores of Higgenbotham students were eighteen months to two years behind those of students at a nearby largely White school and that those differences were perpetuated by the system of curriculum differentiation that existed in the high schools. They were also troubled by the claim that such problems were the result of Black cultural deprivation. The real reason for such achievement gaps, they argued, was that Detroit offered a "dual system" of education that undermined Black achievement.[53]

Three months earlier, the Ad Hoc Committee of Citizens Concerned with Equal Educational Opportunity questioned the apparent assumption of many Detroit educators that the low achievement of Black youth was the result of cultural deprivation. Addressing the poor reading performance of Black students on the Iowa Test of Basic Skills, the committee, which was organized the previous year by a group of prominent Black community leaders, refused to accept the position voiced by many school administrators that saw this problem as the result of the deficient home and community experiences of these students. Such an explanation, they claimed, was an "excuse for denying the potential of the Negro child in our educational system, and to frustrate his efforts to improve his circumstances."[54]

As one member of the committee, Charles Wells, pointed out after listening to countless school administrators voice this viewpoint, "...it became more and more evident that they were convinced that Negro children could not learn beyond a minimal level and that reading programs would reflect this expectation."[55] A year later, the committee challenged the efficacy of the notion of compensatory education that Detroit educators had routinely invoked to justify their reliance on a differentiated, remedial curriculum as the solution to the problems of Black school achievement. They argued that these so-called compensatory programs did nothing to improve the performance of African American students. On the contrary, they claimed that their use had driven Whites out of schools as well as lowered even further the academic achievement of Black students.[56]

In a presentation to the Detroit High School Study Commission, which had been appointed following the Northern walkout to investigate the state of education in the city's high schools, June Shagaloff, Director of Educational Programs for the NAACP, attacked both the practice of curriculum differentiation and the cultural deprivation label. She argued that "for the children who are in the lowest groupings, the children who are the underachievers, all too often instead of getting the program or the kind of educational help that would enable them to catch up, they're given up on and the teachers and administrators who are responsible for this lowest grouping in a tracking system say the best thing we can do is just keep discipline."[57] As a consequence, she felt that a separate vocational curriculum existing apart from the academic program was in effect "a dumping ground for minority group students and for many White students who come from poor families."[58]

Shagaloff also devoted attention in her presentation to what she saw as another myth affecting Black youth, the belief that their low school achievement was the result of cultural deprivation or disadvantage. This was a viewpoint, she argued, that was much more acceptable today than earlier claims to the effect that Blacks were inherently inferior to Whites.

To her way of thinking, however, it was "one of the most racist and class myths yet to develop in America." Its effect, she believed, was to blame everybody but the schools for the underachievement of minorities and the poor while paying no attention to the fact that schools often did little to teach such children.[59]

VI.

This division of opinion among the city's African American population was, however, not to last. Over the course of the next three years, the entry into this conflict of those espousing a Black Nationalist position had the effect of mitigating these internal differences. Black Nationalism was one of the oldest of an array of contending political ideologies that African Americans have advanced to describe their vision of freedom and equality. It traces its roots in Detroit first to the 1920s and the establishment of a chapter of Marcus Garvey's United Negro Improvement Association and a decade later to the emergence of a fledging Black Muslim movement under the leadership of Elijah Muhammad.[60] From the start, Black Nationalists doubted that Whites could be trusted to promote the cause of Black equality and consequently eschewed alliances with them.

Not holding racial integration as a high goal, they were much more interested in securing control of the institutions that served Blacks. They challenged those moderate Blacks who allied themselves with White liberals, who favored racial integration, and who sought to attain access to existing political and economic institutions and to make them work for African Americans.[61]

We can get a sense of what distinguished the city's Black Nationalist educational agenda from the positions taken by other African Americans in Detroit by looking at a series of articles that Karl Gregory, Principal of the Freedom School, wrote for the *Michigan Chronicle*, Detroit's principal Black newspaper, in the four months following the Northern boycott. A good portion of what he said were common complaints that Black Detroiters had made for years and involved finances and segregation. Detroit's schools, he noted, were not receiving sufficient financial support from the State of Michigan to provide a quality educational program. He went on to say that residential segregation exacerbated this problem by leaving the city with a student population whose low achievement made them particularly expensive to educate.[62]

Some of his complaints, however, changed the terms of the debate. Black citizens, he argued, could not trust their schools or rely on the honesty and integrity of the White educators who managed them. He claimed that the school system had not been providing Black parents with the information that they required to recognize the failings of the city's schools. "We cannot be free," he noted in this vein, "until publicly subsidized concealment and misinformation ceases to be the lot of the voter."[63] He went on to say that the difficulties of the city's schools were the result of the fact that Blacks controlled neither their schools nor their community. "People needn't be shocked," he noted, "when residents of a community ask for changes in policies at their school. If students and parents are offended by police practices at a school and an apparent lack of effort on the part of the administration to promote quality education, they have not only the right but the responsibility to demand changes." In a democracy, he asserted, parents and students "must have a share in running their school."[64] As Gregory saw it, then, the problems facing Detroit schools were not simply to be remedied by more money and new laws. Rather, he seemed to be saying that the school system itself was organizationally flawed. The district's leadership was inherently dishonest and routinely lied to its Black clientele, who in turn were shut out from any influence on the workings of the system.

The direction that Black Nationalists were heading becomes clear when we look at a debate that took place at a May, 1966 meeting of Detroit's 17th District Young Democrats involving Carl Marburger, assistant superintendent of schools; Rev. Albert Cleage, a leading Black civil rights leader and community activist; and representatives from Detroit's two teacher organizations. Marburger saw the key problem as that of a lack of money. Detroit, he noted, would have to double its budget to provide its students with the quality of education that was being offered in suburban schools at a time when its tax base was shrinking. A new system of school financing was, he argued, necessary. Patrick Basile, Executive Secretary of the DEA took a similar position. The taxes that Detroit was able to assess were far lower than those of other districts in the state, which made it impossible for the school system to make needed building repairs, adequately fund the teacher retirement system, and hire the best teachers.

Sophie McGlobin, a teacher and a DFT bargaining agent, pointed to the problem of overcrowded classes, which she attributed to a lack of money. She also mentioned non-monetary problems including inadequately trained teachers and poor communication between school officials on the one hand and parents and students on the other. Marburger also called for several non-monetary changes to make the district's bureaucracy more responsive and to alter certification regulations to make more teachers available to city schools. Cleage, however, saw the problem differently. It was, he argued, unlikely that the city's tax base could be improved. He noted that "thousands of White families" were leaving Detroit because they did not want their children to attend inadequate schools. He also blamed the school administration and the city's press for not working hard enough to pass needed bond issues. Key to the deficiencies of the city's schools, he noted, was that "not one Negro is in a policy-making position in the schools or in the city."[65]

Cleage's position, which he developed further in the two years following the walkout, was that Black children could not be educated in Detroit's schools as long as those schools were under White control. Writing in October of 1967, in the *Michigan Chronicle*, about a student protest at Knudsen Junior High School, he stated that many Whites and even some Blacks blamed this protest on the students who they believed were "inferior" and not capable of being educated. It was not surprising, he noted, that there would be African Americans who subscribed to this view. There were those in the Black community, according to Cleage, who have been duped into believing that the school system was trying to provide a quality education and did not realize "that their children are two to three grades behind White children in every subject and that boys and girls are being sent into the ninth and tenth grade who can't even read at a second grade level."

Student boycotts and other disruptions, he argued, did not mean that there was something wrong with Black children. "All that it means," according to Cleage, "is that they are sick and tired of going to schools where they are not learning anything. It means that they can't be expected to respect White teachers who have contempt and fear in their hearts for them. These explosions signify one thing—that there will be no education for Black children until the Black community controls its own schools."[66] Black parents, Cleage argued a month later in the *Michigan* Chronicle; have to recognize that the fault is not theirs or their children's. The problem lay with the White teachers and administrators who have created a school system that was not designed to educate Black youth.[67]

For Cleage, the array of problems facing Detroit's schools, overcrowding, the presence of unqualified teachers and administrators, low achievement, and racism, were the result of the fact that the city's schools were not accountable to its Black citizens. And this was a situation that could only be overcome if African Americans controlled the schools.[68] Black America, he argued, was in the midst of a "Cultural Revolution" that offered African Americans a new sense of identity. They were increasingly becoming aware of their Blackness and their separateness from White society. Blacks looked to the schools to provide their children with the knowledge of African history and the realization that Black science, religion, and philosophy had reached maturity when "the White man was a naked savage living in caves and eating raw meat." In this vein, Cleage believed that Black teachers had a special responsibility. If African American children, he noted "...go to school with a sense of Black consciousness and Black pride, hungrily seeking after understanding of self, and you tell them about George Washington, Abraham Lincoln and the plantation days, you are betraying the trust that Black people have in you." Detroit's schools, Cleage maintained, had to be staffed by Black teachers and administrators. Beyond that, however, the schools had to serve Detroit's Black community, which required Black control.[69]

This was virtually the same message that was conveyed by the Declaration of Black Teachers, which was adopted in April of 1968, at Detroit's Black Ministers-Teachers Conference. The city's schools, according to this statement, were not at present designed to benefit Black youth. For that to happen teachers would have to follow certain "commandments," including:

> We shall not kill the minds and bodies of our children with underestimation of their worth and the worth of Black people.
>
> We shall not adulterate our instruction but shall enrich it with the aim of developing Black youth who will be of service to the Black Community.
>
> We shall not steal their time and energies in busy work in activities designed to promote White class, White values, and goals.
>
> We shall not bear witness against our children nor against our fellow Black teachers but shall do our best to lift them from the hell of ignorance, confusion and despair in which a racist society has placed them.[70]

Black Detroiters who promoted a Nationalist agenda often advanced their case using a politicized rhetoric that masked their educational concerns. A 1969 boycott at Northern High School offers a good example. Like the boycott three years earlier, students, albeit only 125, staged a walkout and sought to establish another Freedom School at St. Joseph's Church. There were, however, some key differences between the two protests. The 1969 protesters emphasized the political nature of their boycott by labeling it as a "revolt." In addition, the demands that the original Northern protesters made for curricular and personnel changes to enhance the school's academic program were less prominent in this protest. Featured instead were demands for the flying of the Unity Flag of the Black Nationalist movement on one of the school's flag poll, the renaming of a reading room in the school in honor of Malcolm X, the introduction of a Black studies curriculum, and the replacement of pictures of White luminaries throughout the school building with pictures of Malcolm X, Martin Luther King, Jr., H. Rap Brown, and Stokely Carmichael.[71]

Yet, Detroit's Black Nationalists did have a distinct educational vision. In their June 1967 recommendations for quality education, the Inner City

Parents Council argued that the low academic achievement of Black youth resulted from the city's schools "deliberate and systematic destruction of the Afro-American child's self-image and racial pride."[72] They went on to claim that in rejecting integration in favor of separation, Black Detroiters were in effect reasserting "pride in their own history, culture, and power."[73] The curriculum was one of the means for securing that separation. Black children, they noted, required a "different educational orientation" than that provided to White children. The course of study had to offer Black students "a knowledge of their history, their culture, and their destiny." The textbooks had to emphasize the worth and value of Black people. Courses should be problem-centered and afford Black children the opportunity to discuss the array of issues that faced their community. And finally, Black children needed a school program that would nurture their artistic and creative abilities. The report concluded with the proviso that all of this must be undertaken without undermining students' academic studies.[74] Although the Parents Council proposed a different curriculum for Black children, it was not a program like Detroit's existing differentiated course of study that would channel different groups of children to different and unequal school programs and ultimate life destinies.

VII.

Educational opportunity was clearly at the heart of the Northern walkout. A week before the boycott ended, Michael Batchelor, one of the protest's student leaders, told the *Detroit Free Press* that "right now the Negro needs an education, and he needs it bad." Another student leader, Judy Walker, told the *Free Press* that she was "willing to give up everything to fight for an education. If we don't have an education, we don't have anything."[75] There was, at least initially, little in the way of agreement among Blacks about the shape that such an education should take.

The remarks of Northern parents and community residents concerning the school's curriculum at the public hearings immediately following the boycott illustrate this disagreement. Some of those in attendance called for a more academically oriented school program noting that more French should be offered, that too many students were being placed in the general curriculum, that more homework should be given, and that a diploma from Northern should qualify a student for admission to college. Others thought that the emphasis should be on vocational training. One parent complained that her daughter had to take a business course after high school. "Shouldn't high school," she asked, "qualify a person for a decent job?" There were suggestions that more work in typing was needed, that the school should focus on preparing "productive citizens" and that Northern should work more closely with business and industry in preparing students for jobs. And still others thought that the curriculum should be more relevant to the lives of Black children. Northern was criticized for teaching less "Negro history" than was taught in Southern schools and for offering a music curriculum that was "White" and out of date. One commentator suggested that students needed a history text that included more "Negro history."[76]

This pattern of disagreement was not unique to Northern. Two years later, a committee of parents and teachers who were investigating community concerns following a walkout at Post Junior High School reported a similar division to that existing at Northern over matters of curriculum. While some community members called for increases in remedial programs, others advocated an upgrading of vocational programs, and still others wanted an emphasis on African American history and culture. And while some community members called for more attention to the honors program, others felt that there should be a focus of attention on courses that addressed the "great social concerns of the day," especially racism, sex education, and drug dependence.[77]

Detroit's Black Nationalists were part of this debate over the education of the city's African American children. Neither willing to accept the existing differentiated curriculum nor a traditional academic course of study, they advocated a school program that addressed what they saw as the unique needs of Black youth in a largely White society. Their proposal called for a school program that acknowledged a rich and vibrant African cultural heritage on which American Blacks could draw while at the same time recognizing their contributions to American civilization. It was an effort that sought to enhance the independence and self-esteem of Black youth. Yet, its presence did not detract the debate's focus from the question of academic achievement. On the contrary, it enriched those discussions by offering the city's Black community another educational vision. This was a vision that both challenged Detroit's longstanding approach to differentiating the curriculum while at the same time providing a pathway different from a White designed academic course of study for enhancing the achievement of Black youth.

These Nationalists, however, did not just contribute one more viewpoint to the array of differing opinions that Black Detroiters held about the education of their children. Their entry into the debate changed its terms. Although never constituting a majority of Black opinion in the city, they were able for a time to solidify an array Black views into a more monolithic oppositional perspective that involved a very different educational vision from that held by the city's White school leaders. They rejected integration and advocated in its place community control. At the same time, Black Detroiters discarded the curriculum continuum that White school leaders had offered them, one that ranged from an academic and preparatory curriculum through vocationalism to remediation. In its place they chose a course of study that infused an academic curriculum with content about African history and culture. No longer were the curriculum and organizational policies about which Black Detroiters and White educators arguing part of the same realm of discourse. They were different and distinct. Writing about a similar conflict between Black and White New Yorkers involving community control of the schools in the Ocean Hill-Brownsville section of that city in 1968, Jerald Podair described that struggle as one between two parties "who spoke different languages to each other."[78] In the aftermath of the Northern walkout, the same could be said about Detroit.

For Black Detroiters, the Northern High School walkout was a concrete and clear indication of their discriminatory treatment, in this case at the hands of the city's White educational establishment. It was an instance in which their experiences of discrimination and subordination became linked to the place where they lived, their shared history, and racial identity to produce a sense of community that was framed in something approach a collective understanding toward the important question of how their children would be educated.[79]

VIII.

In this chapter I looked at the student walkout at Detroit's Northern High School as an example of how Black-White racial conflict that is typically thought as undermining any sense of community had the exact opposite effect. My focus was on how this student protest and subsequent events, particularly the involvement in the conflict of proponents of Black Nationalism, created a more solidified educational vision among Detroit's African American population that challenged the viewpoint of the city's largely White corps of school administrators and teachers. What was happening here was the emergence of a sense of identity among Black Detroiters in reaction to the threats they believed were posed by what they saw as the discriminatory and inadequate education that their children were then receiving in the city's schools. As Cornell West has noted in a conversation with anthropologist Jorge Klor de Alva and journalist Earl Shorris:

> Identity has to do with protection, association, and recognition. People identify themselves in certain ways in order to protect their bodies, their labor, their communities, their way of life; in order to be associated with people who ascribe value to them, who take them seriously, who respect them, and for purposes of recognition to be acknowledged, to feel as if one actually belongs to a group a clan, a tribe, a community.[80]

When a group is able to create that sense of common purpose, the political philosopher Elizabeth Fraser has argued, something important has taken place. They have moved beyond a level of oppression in which "its members share only misery" to a point in which they are able to engage "in the kind of productive argumentative sharing that such groups can sometimes achieve—where the little there is stretches further."[81] In this instance that stretching led to the formation among the city's African American population of a sense of a Black Community. Just as they have for other Americans such efforts at community building have been the ways in which Blacks have identified who they are as a people and what they cherish.[82] As de Alva has noted in the conversation cited above, "no one is born black. People are born with different pigmentation; people are born with different physical characteristics, no question about that. But you have to learn to be black."[83] The Northern High School walkout offers one moment in time where we can see how the kind of learning that creates identity and shapes a sense of community occurred in the wake of the conflict between Black Detroiters and White educators over the schooling provided to the city's African American children.

In the next chapter we jump ahead to Detroit at the end of the twentieth century to view another urban educational reform initiative through the lens of community. The almost forty years between the mid-1960s and the end of the century was a period of flux, conflict, and chaos as the schools and the city itself attempted to cope with the economic and demographic transformations that we described at the beginning of this chapter. There was the continuing racial conflict marked most strikingly during these years by the Detroit riots and the bitter controversy over integration and school busing. There were financial difficulties as the district attempted to cope with enrollment increases and union demands for higher teachers' salaries and improved working conditions with insufficient tax revenues resulting from declining property values and the unwillingness of the city's white population to support tax increases. There were a series of failed organizational efforts that moved the schools from a centralized system to a decentralized system and back again. And there was the continuing inability of the city's schools to address the persistent low academic achievement among African American youth.[84] It was this latter problem occurring as it did as the city's population and political and educational leadership became virtually all Black that would set the stage for a new reform initiative—the mayoral takeover of the board of education.

Chapter 4

Race and Community in a Black Led City: The Case of Detroit and the Mayoral Takeover of the Board of Education

Speaking before the Michigan State Legislature in his 1999 State of the State address, Republican Governor John Engler called for a state takeover of failing school districts. Basing his proposal on the 1995 action of the Illinois Legislature to give Chicago's Mayor, Richard Daley, the power to appoint the city's school board and the system's chief executive officer (CEO), Engler proposed that the Michigan Legislature give the state's mayors authority over their city's schools. In his words, they should have the power to "break the bureaucracy, fix the schools, and put our children first." According to Engler, Chicago's effort, which he claimed enjoyed both bipartisan and multiracial support, was having the effect of improving both the management of that city's schools as well as the educational achievement of its students.[1]

Engler's proposal did represent something of a departure from the long-standing practice that Progressive era reformers had introduced during the early years of the twentieth century of removing education from the control of city government and its then thought to be corrupting influences. At the same time, however, there were counter trends of bringing the administration of schools closer to the city's political apparatus as exemplified by the mayoral control of schools in Chicago, Baltimore, and Cleveland.[2]

As it turned out, however, Michigan during the last ten years seemed to be moving in the direction of greater political involvement in the affairs of local schools. In the late 1980s state officials had in fact threatened the Detroit Board of Education with a takeover if they could not resolve the district's financial problems. In 1989, Mayor Coleman Young had called for the abolition of the board of education and direct mayoral control as a solution to the school system's fiscal difficulties.[3] Five years later, the passage of Proposition A, which shifted the support of public schooling from local property taxes to state appropriations, created conditions for a greater role for the state in the management of public education.[4] Engler had, as it turned out, made similar proposals in the past. In 1996, he had offered Detroit's Mayor Dennis Archer the prospect of managing the city's schools, but Archer turned him down.[5] And in his 1997 State of the State address, he had proposed the state takeover of school districts with high dropout rates or high failure rates on state proficiency tests. In that talk, he made specific reference to two low performing, largely Black school systems, Benton Harbor and Detroit.[6]

In 1999, however, the Republican dominated State Legislature would embrace his proposal. The result was a year and a half struggle that pitted segments of the city's majority Black population against each other. It also brought Detroit's largely Black Democratic legislative delegation into conflict with both the city's African American, Democratic Mayor, Dennis Archer, and the virtually all White Republican dominated state legislature, and created a working alliance around school reform between Mayor Archer and Governor Engler. Although this reform with its mayoral appointed school board and the CEO they selected would only last five years, the conflict surrounding its establishment and implementation offers us an instance in which to explore the role that a concept of community can play in interpreting urban school reform. In chapter one, I suggested the link between community and school organization when I explored the origins and development of community schools. In this chapter, I will consider a similar organizational issue, the mayoral takeover. I will examine the debates surrounding the takeover legislation from the end of 1998 through the decision of Detroiters in 2004 to return to an elected board of education, explore the resulting pattern of agreement and discord, and consider how a concept of community can serve as a lens for interpreting this instance of urban school reform.[7]

II.

Hoping to counter Engler's plan even before it was officially presented to the legislature, the Detroit Board of Education announced some three days before the State of the State address that they would soon be issuing a series of reforms that they labeled as "revolutionary" and "unprecedented." It was a series of changes, according to one board member, Juan Jose Martinez, that would remind people of the reforms that were currently occurring in the Chicago Public Schools. Board of Education President Darryl Redmond challenged the need for a state takeover noting that the district was currently enjoying a budget surplus and that scores on state competency tests were on the increase. As Redmond put it, "we're going to shake up the status quo and make sure that never again anyone will be able to point a finger and say, 'there's bad education in the city of Detroit'."[8]

The following day, Redmond went on to criticize Engler on the grounds that his real motive for the proposed takeover was not student achievement but rather the money that he claimed the state would control from the city's 1994 $1.5 billion bond referendum.[9] Democratic State Senator Burton Leland, a White member of the Detroit delegation in the legislature, viewed Engler's proposal as having "troubling racial overtones." He likened the takeover to the successful effort of the Governor in 1995 to merge Detroit's Recorders Court into the Wayne County Circuit Court. It was a move that many Black Detroiters viewed as an attempt to dilute Black voting strength and the influence of the court's African American judges as well as retribution for the conviction in Recorders Court of two White policemen for the beating death of an African American Detroiter, Malice Green. As Leland stated, "the notion of a White legislature and White governor dictating to a city that is 80 percent African American and a school district that is 90 percent African American is... ridiculous and outrageous."[10]

III.

In February of 1999, four members of the Michigan Senate, including the Republican Majority Leader, Dan DeGrow, proposed legislation to allow mayors in school districts enrolling at least 100,000 students to appoint a five member reform school board and suspend the duties and power of the district's elected board and its officers. The major responsibility of these reform boards was to hire a CEO who would serve at their pleasure. These reform boards were to remain in existence for five years at which time the city's voters could petition for a referendum on their continuation. Although the legislation was written to apply to any Michigan school district, its effective target was Detroit, the only school system in the state with the requisite student population.[11]

Two justifications were offered for the mayoral takeover. First was the fact that Detroit's schools were failing academically as evidenced in low achievement, low test scores, low graduation rates, and high dropout rates.[12] The second reason for the proposed takeover was the widely held view that the city's schools, especially its financial affairs, were poorly managed. It was claimed that the board of education was more interested in providing "perks to its members than on offering its students enough textbooks pencils, and other supplies."[13]

An exchange between Senator DeGrow and the newly elected President of the board of education, Darryl Redmond, that appeared in the *Detroit News* following the introduction of the takeover legislation points to what was an emerging conflict. Invoking the language of the 1983 *A Nation at Risk* report, DeGrow described Detroit's schools as the victim of an "act of war." In this vein, he noted that the city's schools had a graduation rate of 30 percent, that only 6 percent of Detroit's high school students met or exceeded state standards on the High School Proficiency Test, and that only 25 of the city's 245 elementary schools were fully accredited by the state. This was the case, he went on to say, despite the fact that the school system ranked in the top 10 percent in level of expenditures for Michigan school districts and had enjoyed an 18 percent increase in per pupil expenditures since 1967. In this context, DeGrow characterized the takeover bill as an opportunity to reverse this pattern of failure. The proposed legislation was not, he noted, "about process, power, or political agendas. It was about what is best for the children attending the Detroit Public Schools."[14]

Redmond painted a distinctly different picture of the condition of the city's schools. The key phrase in his description was "improved significantly." From his perspective, the schools were "a few years ago" in difficult straits both in terms of the academic performance of their students and the management of their finances. At present, however, the district under the leadership of the board was addressing these problems. He noted that state proficiency test scores have increased, that the city outscored 146 districts in fourth grade mathematics, 231 districts in fifth grade science, and 100 districts in eighth grade writing on state proficiency tests, and that the budget deficit has been replaced by a $93 million surplus. Not surprisingly, Redmond offered a different twist on the issue of accreditation. Ignoring the distinction that DeGrow evidently had made between schools that had received only interim accreditation and those better performing schools that were fully accredited, he claimed that all of Detroit's schools were in fact accredited. Redmond also noted that Detroit did not have the academic and budgetary problems of the Chicago Public Schools, which seemed to be the model that supporters of the takeover used to justify their efforts.

In addition to challenging the actual assessment of takeover supporters, Redmond injected another issue into the debate, that of local control. Not only did the takeover legislation, in his estimation, fail to address the problems facing the city's schools, it did not "create the conducive collaborative atmosphere for all stakeholders." He went on to say "this legislation seemingly was created in a vacuum by state bureaucrats in the executive branch without consultation by superintendents, parents or other public school stakeholders."[15]

A few days earlier, Detroit Superintendent Eddie Green had also voiced opposition to the takeover. Like Redmond, he noted that Detroit students need to perform better than they are doing but that that they have made significant progress during the last four years. He claimed in that vein that Detroit children were currently outscoring such districts as Grand Rapids, Flint, Lansing, Pontiac, and Muskegon in some areas of the state proficiency tests. Green also wondered why there was such a rush on the part of the state to replace the city's board of education. He noted that the board would soon reveal its own reform plan and that what was called for was a thorough discussion and debate on various proposals for improvement that involved all of the "stakeholders." Finally, Green questioned why Detroit was singled out for a takeover. The initial version of the legislation allowed for a state takeover of any school district that did not meet certain academic and fiscal conditions. He charged, however, that as a result of political pressure, the final version of the legislation was limited to Detroit.[16]

In mid-February in response to the takeover legislation, the board of education issued its own reform plan. First, parents would be asked to volunteer for forty hours during the year to help in their children's school, either in the classroom or on the playground. To increase the likelihood of this happening, employers and unions would be asked to allow parents to take a half a day a month to work as a volunteer in their children's school, and clergy would be asked to promote such participation from the pulpit. As part of this provision, parents would be required to assist the school in managing any discipline problems exhibited by their children. Second, the board of education would resist efforts to micromanage the school system. The authority to approve bids and issue contracts would be shifted from the board to the superintendent. Third, the board would use all of its current budget surplus to address the hundred or so lowest performing schools in the city. Finally, the board agreed to implement an array of management reforms that had been recommended by New Detroit, a coalition of White corporate executives and Black community leaders dedicated to civic improvement and improved racial relations.[17]

The hearings before a joint meeting of the House and Senate Education Committees during February brought out an array of opponents and supporters of the takeover legislation. Some of the opponents echoed Senator Leland's charge that the motivation behind the proposal was racial. Democratic Representative Ed Vaughn, an African American from Detroit, challenged Senator DeGrow's assertion that the city was singled out for this legislation because of its size. He claimed that the takeover was primarily about race. "The only school district targeted," he charged, "is the biggest and the Blackest." More common among the opponents, however, was the claim that the takeover legislation violated both the principles of local control and threatened the voting rights of Detroiters. Rev. Thomas Jackson, a Baptist minister whose six children had attended the city's schools, commented at a citizens' roundtable organized by the *Detroit News* that "everybody that's voting on this thing to take way my right to vote are [legislators from] another area, and I don't have the right to vote them out of office." This interpretation of the takeover effort was not, however, devoid of racial overtones. Jackson went on to say that this initiative was both "wrong" and "racist" because those who were voting on the mayoral takeover were not "beholden" to him.[18]

Supporters of the takeover expressed a quite different view. Some, like Richard Blouse, President of Detroit's Regional Chamber of Commerce, noted that the legislation is supportive of local control by placing the power to appoint the board of education in the hands of Detroit's mayor. The Republican Chair of the Senate Education Committee, Loren Bennett, went further and claimed that local control was beside the point. Local governments, he noted, derived their authority from the state, and the state could enact legislation to alter that authority.[19]

Popular opinion more or less paralleled the divisions among those attending the hearings. A February 14 poll published in the *Detroit News* reported that a majority of Detroiters believed that the schools were in need of reform. When it came to the takeover legislation, however, about three quarters of Whites were in support of the proposal while over half of the Blacks polled were opposed. Sterling Jones, a Black carpenter who had a stepdaughter in the city's schools, saw the takeover as "a racist move." "The Republicans," he stated, "are primarily a gang of White men trying to do whatever they can." Angela Williams, an African American mother of four stepchildren in the city's schools, stated that she was aware of the problems facing Detroit's schools but was afraid of the prospect of losing her vote if the takeover proposal was successful. On the other side, Frances Irwin, a White parent who has adopted two sons, one Black and one White, lauded what she saw as Engler's effort to do something about the condition of Detroit's schools.[20] In another sampling of public opinion a few days earlier, Earl Chambliss, a Black Detroiter who sent his son to a parochial school, commented to the *News* that he has lost confidence in the city's schools.[21]

IV.

To move the takeover forward, Engler looked to Detroit's Mayor, Dennis Archer, for help. He wanted Archer to be ready to appoint a reform board as soon as the legislation passed and to have a new CEO selected by July.[22] The previous December, Archer, who was Detroit's second Black mayor, following Coleman Young into office in 1993, appeared to be a likely supporter of a takeover when he told the Detroit Board of Education to "shape up or else."[23] Yet once the actual takeover legislation had been introduced and was being debated, his position was harder to pin down.

For several weeks, Archer claimed that he had not made up his mind regarding the takeover bill or his role in the effort.[24] In his State of the City Address in February of 1999, Archer urged caution about the takeover proposal and called for its careful scrutiny before any plan was enacted. He went on to say that the kind of legislation that he wanted would do more than just provide the mayor with power to appoint the board of education. He thought that there were a number of other initiatives that should be part of this legislation, including a cap on class size of 17 in grades one to three and 20 in the remaining grades, mandatory summer school for underachieving children, the establishment of after school programs, and a program for recruiting 1200 new teachers for the city. What the city schools needed, he noted, was a "Marshall Plan." Finally, he urged Senate Majority Leader DeGrow not to report the takeover bill out of committee until Archer had the opportunity to share his views with the legislature.[25]

A week later, at a speech to the Economic Club of Detroit, Archer took a more definitive stand in favor of the takeover and stated if the legislation passed, he would, in his words, "take the bull by the horns." He went on to say that he was currently working with the legislature to incorporate into the bill some of the provisions he called for in his State of the City Address.[26] What seems to have brought Archer on board was a bipartisan deal negotiated by Senator Virgil Smith, the only African American member of the Detroit legislative delegation in support of the takeover, to provide the city's schools with an additional $15 million to help pay for the initiatives that Archer had requested.[27]

Archer's slowness and sometimes equivocation on the takeover can be explained by a number of factors. He was in the midst of his own struggle to fend off a recall drive that began six months after his reelection to a second term at the behest of a coalition upset with his failure to back a proposal to give one the city's new casino licenses to a local Black developer. Although the recall effort would ultimately fail, there were other doubts raised about his leadership ability as a result of a failed effort to deal with snow removal in the wake of a surprise blizzard in January of 1999 that left many of the city's residential streets blocked for almost a week. There was some speculation that the city's business leaders would seek an alternative candidate if Archer sought a third term.[28] And there were city residents who were less concerned that he reform the schools than that he act on an array of what they saw as neighborhood issues, including the modernization of the fire department, community policing, and the demolition of abandoned buildings to name but a few. In this vein, the President of the Bewick Block Club, Leontine Person, told the *Detroit Free Press* that "I want to see him stick a shovel in the ground and say this is what we're going to do for the people of Detroit instead of what these big corporations are doing."[29]

Beyond these immediate issues, Archer's leadership style seemed to frustrate some Detroiters. More of a conciliator and more willing to try to work with Detroit's corporate leaders and White suburban communities than his predecessor, Coleman Young, Archer's penchant for compromise and consensus, which was no doubt due to the fifteen years he spent as a state supreme court judge, was often interpreted as inaction and an inability to lead.[30] Pete Waldmeir, a columnist for the *Detroit News,* claimed that Archer did what he did best in his comments on the takeover in his State of the City Address. In Waldmeir's words, "he talked about it." The columnist went on to question why the mayor had asked Senator DeGrow to delay reporting the bill out of committee until "who knows when." As Waldmeir put, "maybe he thinks lightning will strike and fix everything."[31] The following day another *News* columnist, George Weeks, urged Archer to act with more speed. "Archer should," he noted, "seize the moment, not try to delay it."[32]

While Archer was deciding what role that he would play in the takeover initiative, the legislature was continuing its deliberations. On February 25, 1999, as 100 or so protesters crowded the halls of the capitol in Lansing chanting, singing, and issuing threats of recall, the Senate Education Committee, although interrupted briefly by two Detroit House members who voiced opposition to the bill as an "affront" to the voting rights of their constituents, voted 3-1 to report out the takeover proposal to the full Senate without amendments.[33] At the same time, a number of Senate Democrats were contemplating attempts to amend the bill on the Senate floor. Gary Peters, a Democratic Senator from Bloomfield Township in suburban Detroit, had two such amendments in mind. One would require a referendum by Detroit voters to approve the takeover. He argued that the abolition of the elected board required by the legislation as well as the fact that the mayor had not been elected to operate the schools had implications for the "voting rights" of Detroiters that could only be settled by popular vote. The other amendment that he wanted would require an election at the end of five years to determine if Detroiters wanted the takeover plan to continue. As Peters saw it, the Republican legislature had the votes to approve the takeover legislation for implementation the following year. What they lacked, however, was the two-thirds majority necessary to implement the bill immediately as Governor Engler wished. Under such circumstance, he believed that there was an opportunity for Democrats to play a role in shaping the final outcome.[34] As it turned out, neither amendment ended up in the final bill that the Senate passed at the beginning of March.

Rather than amending the Senate approved bill when they received it, the House with bipartisan support passed its own version of a takeover bill. Under the provisions of this plan, Governor Engler was to appoint a monitor who would run the schools while retaining the elected board of education in an advisory capacity. When the monitor's term expires in 2003, board members could run for re-election and their authority to manage the schools would be returned to them.

The six Detroit Democrats who supported the House takeover plan claimed that it, unlike the Senate bill, protected the voting rights of Detroiters and placed responsibility for reform clearly on the governor. Equally important in their support of the House alternative was their distrust and dislike of Archer. They generally questioned the quality of his leadership of the city as well as being upset that they had not been consulted in the development of the original takeover plan. There was also some feeling among these legislators that Archer's apparent support of the takeover proposal was a betrayal of Black interests. One of their number, Representative Lamar Lemmons, noted that if there had to be a takeover, it would be better to give the power to Engler than to Archer. "If you want a plantation analogy," he told the *Detroit Free Press*, "it's African Americans' experience that overseers are often worse than dealing directly with the master."[35] The House bill was not just, however, an attack on Archer. The minority Floor Leader of the House Democrats, Detroiter Kwame Kilpatrick, saw the effort as a "strategic move" to force Republicans to negotiate on the terms of the takeover.[36]

In response to the House action, the Senate did in fact introduce some changes into their takeover legislation. They altered the membership of the appointed board to seven members, six appointed by Mayor Archer with the seventh seat given to State School Superintendent Art Ellis or his designee. The board would, in turn, appoint a CEO. And they agreed to maintain the elected board of education in an advisory capacity.[37] Other provisions of the act would allow the new CEO to fire teachers and principals, waive provisions of union contracts, and reorganize schools that were viewed as failing. A Week later, both the Senate and the House passed Public Act 10 with sufficient majorities to allow for its immediate implementation.[38] Governor Engler quickly signed the bill into law and Mayor Archer designated Detroit's current General Superintendent of Schools, Eddie Green, as Acting CEO until the new reform board selected a permanent CEO.[39]

V.

The dispute over the takeover legislation involved two distinct but related conflicts, one having to do with the governance of the Detroit Public Schools and the other dealing with political control of the city. The battle over the city school's governance pitted two loose groupings of individuals. On one side stood a Black and White coalition in support of restructuring the schools including Governor Engler, Mayor Archer, a few key Black community leaders, the Republican majority in the state legislature, a handful of Democratic legislatures, the city's two major newspapers and a number of their columnists, Detroit's principal Black newspaper, and, according to one poll, about 40 % of the city's Black population. Arrayed against them was a largely but not exclusively African American alliance comprising the vast majority of Detroit's Democratic legislative delegation, the Detroit City Council, several Black community leaders, a number of politicians, school administrators and board of education members outside of Detroit, and, according to the poll cited above, about half of the city's Black population.[40]

At one level this was clearly an ideological conflict about how the city's schools should be run. Many of the supporters of the takeover legislation invoked the image of Chicago when they talked about the potential that a mayoral takeover held out for Detroit. What is less clear, however, is whether either the supporters or the opponents of the takeover understood where the changes implemented in Chicago fitted into the larger process of school reform that had been underway in the nation since the early 1980s.

When the takeover was first proposed by Governor Engler, Joe Stroud, a columnist for the *Detroit Free Press,* pointed out that reforming the schools would require both "top down" district level reforms like the takeover to improve school accountability along with "bottom-up" initiatives involving parents, teachers, principals, and students to make changes within the classrooms.[41] In a later column, midway through the conflict, he pointed to the liabilities of relying wholly on either kind of initiative. As he saw it, a powerful state bureaucracy could interfere with the efforts of parents and teachers to improve the quality of education in the classroom. Local control, however, could undermine wider efforts to introduce higher standards. What had to be done if Detroit's schools were to improve, he argued, was for reformers to strike something of a balance between centralizing and decentralizing impulses so that the status quo would be challenged with "accountability imposed from above as well as from the empowerment of parents and students at the most intimate level." There were no simple solutions for addressing the problems facing public schools. "Working from both ends, though, from the top and the bottom, we just might get the education results we want for more nearly all the children of Michigan."[42] Stroud, then, seemed to be aware of the shifts between centralization and decentralization that had punctuated the various so-called waves of school reform that were then occurring as well as recognizing the strengths and limitations of the kind of restructuring that a mayoral takeover posed for school governance.[43]

The majority of the proponents of the takeover did not, however, frame their support in the language of educational reform or restructuring. Although both the *Detroit Free Press* and the *Detroit News* supported the takeover on their editorial pages, it was unclear what, if any, reform ideology drove either newspaper. Their advocacy of this initiative seemed to have little to do with their belief in any particular set of educational ideas but rather was motivated by more pragmatic concerns about the ability of the city's schools to operate effectively under its present administrative structure. A takeover was called for, according to both of these newspapers, because of the schools' seeming failure, despite numerous opportunities over a period of years, to educate Detroit's children and the board of education's ineptitude, if not outright corruption, when it came to managing the district's finances.[44] The *Michigan Chronicle*, the state's Black newspaper, saw the takeover as a means of wresting control of the schools from a politically oriented board of education whose major commitment was in maintaining its power. As the editors saw it, the existing board had placed a concern for the education of the city's children behind their concerns with "making a good impression at public meetings, controlling contracts, dispensing favors, and building a base to launch future political campaigns."[45]

Similarly, the comments that ordinary Detroiters made to the press in support of the takeover did not seem to be driven by any specific educational ideology but rather by the belief that their schools were not well managed and not working to educate the city's children.[46] As Yvette Anderson noted in a letter to the *Detroit News*, the takeover battle was not about race. "Black people," she went on to say, "just want a proper education provided to their children so they can become productive citizens."[47]

The opponents of the takeover opposed this initiative on both pragmatic and ideological grounds. As a practical matter, those who were against the takeover thought that it was unnecessary because Detroit's schools were in fact improving on a number of fronts including student achievement. Beyond that, however, they did mount an ideological argument against mayoral control. Such an initiative, they argued, violated the principles of local control, denied Detroiters their voting rights, and was racist in intent. As Rev. Leonard Young, a Baptist minister from Detroit, made clear at the initial hearings before the House and Senate Education Committees, the takeover legislation was an effort to subvert a local election. He went on to say that "if you respect Black people, you will respect our vote."[48]

Although the opponents to the takeover often expressed their descent in racialized terms, it is important to note that this dispute was not a monolithic Black-White conflict. In fact, there were Blacks and Whites on both sides of the issue. The city's two major Black organizations, the NAACP and the Urban League, took opposing positions on the takeover. The NAACP opposed the takeover on the grounds that removing an elected board of education in favor of an appointed one threatened the voting rights of Detroiters. The League, however, supported the takeover. The key issue, as they saw it, was not voting rights but the need to have quality schools that would prepare the city's children for jobs and other opportunities.[49]

Supporters for the takeover, in fact, included Black organizations such as the Detroit Association of Black Organizations, 100 Black Men, and the Ecumenical Ministers Alliance as well as such integrated groups as the Detroit Federation of Teachers, the Detroit Organization of School Administrators and Supervisors, New Detroit, and the Detroit Regional Chamber of Commerce.[50] Rev. Horace Sheffield, pastor of Detroit's New Galilee Missionary Baptist Church and co-chair of the Ecumenical Alliance, reflected the frustration of many of these supporters over the emphasis that had been given to issues of race in this dispute. He challenged the claim of many of the opponents of the takeover that attributed racist motives to Engler and those Whites who supported him. In his words, "the only race that matters is the race that succeeds in properly educating our children before another generation is lost."[51]

Similarly, opposition to the takeover did not just come from Detroit Blacks. A number of largely if not virtually all White school districts, including such suburban Detroit systems as West Bloomfield, Royal Oak, and Hazel Park as well as those in the rural out-state communities of Boyne and L'Anse, opposed the takeover as did a majority of the membership of the Michigan Association of School Boards. The issue for these opponents was not of course race but local control. From their vantage point, the takeover proposal, if passed, would provide a precedent that might threaten the future independence of their districts.[52]

The battle for political control of the city was a conflict within Detroit's Black community involving Mayor Archer on one side and the city's legislative delegation, particularly State Representatives Ed Vaughn and Kwame Kilpatrick on the other. This dispute was certainly connected to the takeover conflict in that educational reform was, as we have seen, a point of division for them. Yet, the quarrel between Archer and these Black Detroit Democrats predated the takeover debate. Ed Vaughn had run for mayor against Archer in 1997 and lost, while both Kilpatrick's parents, Congresswomen Carolyn Kilpatrick and Wayne County Commissioner Bernard Kilpatrick, had at times contemplated running for mayor. Kwame Kilpatrick and his father, in fact, were members of the African American Men's Organization, a group that was organized in 1997 to counter Archer's influence.[53]

The takeover offered Archer's Black rivals the perfect opportunity for expressing their antipathy toward him. Archer's ongoing problems, particularly the recall attempt, his botched effort to deal with snow removal, and the criticism of his conciliatory approach to addressing city problems had rendered him vulnerable to attacks from his opponents.[54] The mayor's key Black supporters in Detroit had their own weaknesses. Bill Beckham, President of New Detroit, who died suddenly in the midst of the takeover battle, was rumored to be a possible challenger to Archer if the Mayor sought a third term. And Senator Virgil Smith's support was attributed to his desire to cooperate with Republicans in order to secure an appointment of some kind when term limit legislation forced him to leave the Senate in 2002.[55] Under these circumstances, the support that Black Detroiters gave to the passage of a House version of a takeover bill was their way of embarrassing Archer by telling him in effect that they preferred the leadership of a White Republican governor to their fellow Black Democrat.[56]

The takeover dispute was, however, not just about conflict. It also was the vehicle for an alliance between a Black, Democratic mayor and a White, Republican governor. Although Archer and Engler voiced similar criticisms of the Detroit Public Schools, it is not clear to what extent they agreed on educational matters. In fact, when the House passed its own takeover proposal that placed the Governor in charge, Engler angered Archer by stating that he would be satisfied with either the Senate's mayoral takeover plan or the House legislation. He did, however, after a brief meeting with Archer following the House's action, reaffirm his support for the Senate plan.[57]

In the end, Engler and Archer's alliance may have had more to do with the politics of the moment than with educational ideology. The success of the takeover proposal was the most recent indication of the fact that Detroit was losing the influence that it once had enjoyed in the state legislature. The city's declining population, Republican control of the state government, and recent term limit legislation were the culprits behind this change. Between 1950 and 1990, Detroit's population fell from about 1.8 million to just over 1 million. As a result, the city had lost membership in the state House of Representatives from thirty seats in 1950 to twenty in 1970, to thirteen in 1999. Declining population had also affected Detroit's voting power. In the 1958 gubernatorial election, about 25 percent of the votes came from Detroit. In 1978, the city accounted for 11.5 percent of the votes, and in the 1998 election for 7.5 percent of the votes. And recently passed term limit legislation would soon force the most senior members of the Detroit legislative delegation, whose tenure and experience provided them with the greatest influence, to leave office.

The result was that Republicans at the state level could act notwithstanding the opposition of Detroit Democrats. In the 1998 legislative session, they were able to pass a bill to mandate drug testing of welfare recipients despite objections from Detroit lawmakers. And they were able to force the city to lower its income tax in exchange for a guaranteed share of state revenues. It was no doubt this shift in power relations that Representative Vaughn had in mind when he accused Governor Engler of practicing, in his words, "plantation politics."[58]

Seen in this light, Engler may have viewed the takeover as a means of cementing an alliance with Archer that would divide state Democrats and, as a result, further diminish Democratic Party power in Michigan. For Engler, the takeover, if it actually served to improve the schools, offered him the opportunity to enhance his reputation as a visionary and a leader without much risk. He could undertake the initiative without worrying about angering Detroit Blacks since he did not need their votes to get elected. And if the takeover failed to improve the schools, he could blame Archer. The Mayor was also attracted to the takeover by the prospect of reforming the city's schools, which would further advance the leadership role that he was assuming in national Democratic Party politics. For Archer, however, the venture was riskier. If the schools did not improve, he would in all likelihood suffer increased opposition to his administration.[59]

VI.

In the same month that the Michigan Legislator passed Public Act 10, a majority of Detroiters, it seemed, were in support of the takeover.[60] Yet, divisions over the reform that were present at the outset did not disappear. The legislation that enacted the takeover required that the reform board be unanimous in selecting its CEO. When the board was not able to agree on a candidate for this position, they appointed David Adamany, former President of Wayne State University, as the school system's interim head. During his short time in office, he instituted a plan that reduced class size, addressed the problem of excessive student truancy, privatized the purchase of school supplies, introduced more efficient procedures for undertaking construction, and began to spend the money that a 1994 bond issue had brought the district.[61]

Where he did, however, encounter difficulty was in his effort to renegotiate the contract with the Detroit Federation of Teachers. Adamany and the union were in agreement on some issues, such as excessive truancy among high school students, the practice of social promotion, the failure of administrators to enforce discipline and attendance policies, and class size. There were, however, proposals that he made that the teachers opposed. They included an extension of the school day without additional payment to teachers, merit pay for high achieving schools, penalties for teachers who used their sick leave for personal absences, and a provision to allow the CEO to reorganize low achieving schools by replacing their teachers and administrators. The result was a short teacher strike in August of 1999 that ended with both sides withdrawing some of their more contentious demands. The board withdrew their proposal for a longer school day and their merit pay plan while the union accepted the CEO's right to reconstitute failing schools with some protections for affected teachers. Under the new agreement, teachers received a 6 percent pay increase during the three-year contract.[62]

Adamany had a contentious relationship with the unions representing faculty while he was President of Wayne State and it was not surprising that he clashed with the Detroit Federation of Teachers. The conflicts that the first permanent CEO, Kenneth Burnley, former Superintendent of the Colorado Springs Public Schools, experienced when he took over in May of 2000 were slow in coming and less expected but deeper in that they resurrected some of the divisions that had existed surrounding the takeover itself.[63] He started off on a strong note with major improvement plans focused largely on the district's administrative efficiency but also including efforts to improve student academic achievement.[64] Although the assessments of his performance at the end of his first year in office in 2001 were quite positive, his efforts to improve the school system's management would create problems for him.[65]

In office, Burnley continued a financial audit of the system that Adamany had begun. The first such audit in twelve years, it revealed numerous instances of shoddy bookkeeping, charges of embezzlement against three former school bookkeepers, and the citing of fifteen high school principals for the misuse of school funds. One of the most extreme of these cases was that of the Principal of Henry Ford High School who claimed reimbursement for numerous personal expenses including alcohol, home improvement expenses, and his son's vacation in Italy. All in all, the auditors recommended that he repay the district over $76,000 for unallowable expenditures.

The fact that the audit did result in the prosecution and firing of many of the school personnel involved in these misappropriations can be seen as pointing to the potential benefits of the takeover. Yet, the audit brought with it criticisms of Burnley that weakened his status as a reformer. As part of his plans to improve student academic performance within the district, Burnley had created new administrative positions with the title of Executive Director of Accountability. Those who were to fill these new roles were charged with providing individual principals with assistance in remedying the district's low graduation rate and low student performance on standardized tests. The audit, however, revealed that four of the former principals that he appointed to fill these new posts had been cited for their misappropriations of school funds. It was also pointed out that seven former principals who were promoted to these new positions had led schools with low test scores that had declined even further under their leadership. Burnley himself, it turns out, was accused of cronyism because among these appointees was his track coach when he was a student at the city's Mumford High School and the wife of one of the members of the reform school board. Not only was Burnley criticized for appointment practices. The *Detroit News* accused him of withholding the results of the financial audit for four months. They went on to claim that he, in fact, knew about the result of the audit before he promoted the four principals cited in the report for financial irregularities.[66] Yet, despite this criticism of Burnley, he was able to complete his second year with a generally positive evaluation by the reform board and a survey of city residents that noted continued support for the takeover initiative.[67]

Questions about the district's financial management did not, however, go away. The ongoing process of auditing the city's schools pointed to continuing instances within the system of inappropriate expenditures and lost money.[68] Ultimately more damaging to the reform board and to Burnley himself was the controversy over the actual success of the takeover in improving the city's schools. In both his 2003 and 2004 State of the District Messages, Burnley acknowledged the financial, management, and educational challenges facing the district but noted improvements in student academic performance.[69] Others, however, were less certain about the degree of improvement. In October of 2004, the *Detroit News* acknowledged that there had been significant improvements in the management of the district and in the quality of its physical plant. It even noted improvement in student performance in some of the lowest achieving schools. Yet, the article reported declines since the takeover in key state test scores in comparison to average scores of other urban districts in the state as well as the inability of the district to increase the high school graduation rate above 50 percent.[70]

Popular opinion about the success of the reform began to change. A majority of respondents to an April 2003 survey of 400 city residents indicated support for abolishing the reform board with over 60 percent of them voicing a negative view of Burnley's performance.[71] This shift in public opinion was due in part to financial problems facing the district as a result of a reduction in state funding and revenue losses accompanying enrollment declines.[72] Burnley's decision to address these problems by closing schools and reducing staff led key groups of parents, teachers, and others to call for his removal.[73] There were suggestions, however, that Burnley got caught in a climate of popular anger and distrust over what some Detroiters saw as an effort of a largely White run state government to take away the voting rights of Black Detroiters.[74] It also might be the case that Kwame Kilpatrick's renewed involvement with the issue of the takeover may have spurred opposition. In 2002, he was elected Detroit's mayor when Archer decided not to seek a third term. Although opposed to the takeover when he served in the Michigan legislature, Kilpatrick stayed clear of educational politics during his first year in office.

In September of 2003, Michigan's Democratic Governor, Jennifer Granholm, proposed a plan in which the mayoral appointed board would be replaced by an elected board whose powers would be limited to appointing and removing the chief executive officer with the mayor having authority to approve or veto the appointment. Her suggestion for an elective board brought Kilpatrick into the debate with yet another alternative plan that called on Detroit voters to decide in March of 2004 on whether to return to the old elected board of education or to establish a new system in which there would be a nine member board elected by districts with a CEO appointed by the mayor. Under his proposal, the new board would monitor student performance, review annual financial audits and the annual budget, and provide the mayor with an annual evaluation of the CEO. The CEO, who would be appointed by the mayor, would retain his authority to manage the day-to-day operations of the school.[75]

It is not certain why Kilpatrick entered into the debate. There were charges made by those opposed to the takeover that he was attempting to increase his own power and at the same time gain control over the city's school budget. At the least doing this would, it was claimed, make him the most powerful mayor in Detroit's history. It would also give him access to new funds that he could redirect toward his own upcoming reelection bid.[76] It may have also been the case that from his vantage point as mayor he saw the importance of having a powerful school chief, but at the same time he wanted to reassert the rights of Detroiters to elect their own board of education.

In any event, the introduction of new proposals by the governor and by Kilpatrick precipitated virtually the same conflict between two largely Black coalitions that the original takeover proposal had brought.[77] On one side stood those who wanted to return to the old elected board with a board appointed superintendent. The quality of education provided to the city's children, according to these opponents, had not improved under the reform board. Beyond that, anything less than a return to the old elected school board would remove popular control of school matters from Detroit citizens, which, as they saw it, was racially motivated.[78] On the other side were those who supported a combination of an elected board and a CEO appointed by the mayor. Denying that Kilpatrick had supported a mayoral appointed chief executive officer for personal gain, they saw this approach as a solution that preserved the voting rights of Detroiters and continued the city's educational reforms. These were reforms, they argued, that would enhance student academic achievement, produce a more skilled workforce, and provide the city with a more competitive economy. They went on to say that a support for the old elected board would mean a return to the managerial inefficiency, corruption, and poor educational performance that had been a longstanding feature of education in Detroit.[79]

Ultimately, Kilpatrick's proposal did not receive the required state legislative approval. A substitute plan that the mayor did endorse was subsequently approved that allowed voters to decide in a November 2004 referendum between a return to the old elected school board or a plan that would establish the nine member elected board that Kilpatrick wanted but limited its power to approving or rejecting the mayor's appointment for CEO.[80] Kilpatrick's proposal for an appointed CEO was, however, defeated. The following year, Detroiters reelected him to a second term as mayor and elected a board of education for the first time since 1999.[81]

VII.

In chapter one I argued that the concept of community is what can be called a floating signifier that, despite its multiple referents, points to the existence of discursive practices that indicate a sense of collective belonging among individuals and groups. Civic capacity, which I also addressed in chapter one, represented another way of indicating the presence of the kind of agreements that points to the existence of community. Those who write on mayoral control often talk about civic capacity as both a prerequisite for mayoral control as well as an outcome of such control.[82] The Detroit takeover, however, brought with it an array of conflicts and divisions. There was the initial struggle among Detroiters over the takeover proposal. There were conflicts between the interim CEO and the teachers union. Once the reform board was up and running and a permanent CEO had been selected, Detroiters were divided about the effectiveness of this new organizational scheme as well as the effectiveness of Burnley's leadership. There was the disagreement about whether to maintain a system of reform or return to the old elected board of education. And finally, throughout the entire life of the reform board there was the suggestion that at root all of this point to a Black-White racial divide. In the previous chapter, I argued that the struggle between Black Detroiters and the Whites who led school system during the 1960s created the conditions that allowed for the emergence of a sense of community among African Americans. Clearly that sense of belonging was not permanent. By the time of the takeover conflict, it was all but dissipated.

In his discussion of the contradictory goals that Americans routinely invoke to talk about the purposes of public schooling, the sociologist David Labaree offers a way of cataloguing this conflict surrounding the takeover. As Labaree sees it, debates about school restructuring, which were at the heart of the Detroit takeover battle, are one of the myriads of issues about which educational reformers routinely argue. Such a conflict, he goes on to say, is not really about organizational issues but about political ones. That is, the dispute over the takeover was not so much a battle over differing curricular and pedagogical ideas but rather about the goals to which Detroiters wished their schools to be directed.

Labaree identifies three such conflicting goals—democratic equality, social efficiency, and social mobility—that have emerged out of the historic tensions and contradictions in American society between the ideals of democracy and those of capitalism. Arrayed against each other for the loyalties of Americans, it is the ebb and flow of their influence as they come to the fore, retreat to the sidelines, and then emerge again that characterizes for him the history of American education. From the perspective of democratic equality, education is a public good that benefits all members of society by preparing youth with the knowledge, skills, and dispositions necessary for citizenship and political participation in American society. To ensure that all students have equal opportunity to acquire these educational opportunities, this goal requires that schooling be widely accessible to all segments of society. A second goal for Labaree is that of social efficiency. Here, too, education is a public good for promoting society's economic development and well-being. The purpose of schooling from this perspective is to prepare youth to fulfill the economic roles that are essential for a prosperous and productive economy. Meeting such a purpose requires a differentiated curriculum that channels students toward their appropriate occupational destiny, which in turn entails that schools will be highly stratified with selective access. A third goal that Labaree identifies is that of social mobility. Education, according to this viewpoint, is a private good that enables individuals to obtain an advantage over other members of society. The goal of education from this vantage point is to prepare youth for desirable social, economic, or political positions and in so doing to offer them a competitive advantage over others.

Labaree's framework does indicate a clear political conflict between the supporters and opponents of the takeover. In defending a mayoral takeover, its proponents talked at times about the potential of this reform for enhancing standards, improving test scores and graduation rates, and eliminating waste in the administration of the schools. Such results were important for these reformers because they would enhance the efficiency of Detroit's schools in preparing its students for their occupational roles in adult society. In promoting the takeover on these grounds, its supporters were in effect extolling the goal of social efficiency. The opponents of the takeover, on the other hand, challenged this proposal on egalitarian grounds. As they saw it, the takeover threatened the voting rights of Detroiters thereby undermining their ability to act as participating citizens in a democratic society. Such voting rights were of particular salience for African American they went on to say, because they were the hard fought fruits of their victory over the forces of discrimination and racism. In making this argument, the opponents of the takeover were, in effect, justifying their position in terms of the goal of democratic equality. A similar division of opinion pervaded the discussions surrounding the success of the reform board and its future.

Lest we assume, however, that the positions of the supporters and opponents of the takeover were that unified and monolithic and the battle lines between them were that clear, we should also note a commonality in their arguments to which Labaree's framework points. In defending their positions, both supporters and opponents argued about the relative success of Detroit's schools in educating children in general and their children in particular. That is, when the supporters of the takeover talked about the failure of the city's schools, they were, in effect, noting the fact that these schools did not offer children, particularly Black children, the knowledge and skills that they needed to compete effectively in adult society. And when the opponents of the takeover noted the fact that Detroit's schools were improving, they were claiming that these schools were making progress in offering the city's children the competitive advantage that they required to succeed in the larger society. To the degree that both the proponents and opponents defended their position in these terms, they were invoking the goal of social mobility.

Labaree alerts us to the fact that it was not unusual for both the supporters and the opponents of the takeover to share a commitment to the goal of social mobility. Proponents of democratic equality have often invoked the doctrine of social mobility to bolster their opposition to the goal of social efficiency. Both democratic equality and social mobility, he notes, can be invoked to justify increasing student access to educational opportunity over and against the tendency of advocates of social efficiency to favor differential and restricted access to schooling. Similarly, supporters of social efficiency have often favored the goal of social mobility to strengthen their opposition to democratic equality. Labaree points out in this vein that both goals can support the commitment to increased stratification through tracking and curriculum differentiation as opposed to the challenge that a belief in democratic equality poses to these two practices.

Looking at the takeover from the vantage point of Labaree's framework does suggest how a notion of community can help us in interpreting this instance of reform. The conflict between the supporters and opponents of the takeover, using Labaree's terminology, was a struggle between those who saw education as a vehicle for democratic equality and those who saw it as a means for promoting social efficiency. Both of these viewpoints were framed in what we might think of as a communitarian discourse that depicted Detroiters as being bound together into a collectivity whose purpose was to promote a common good. Where they differed was in their understanding of what that common good entailed. It could, as it did for the supporters of the takeover, point to the role that the city's schools could play in building the skilled and knowledgeable workforce that offered Detroit an efficient and productive economy. Or it could lead Detroiters to oppose the takeover on the grounds that such a reorganization of its schools violated the principles of democracy and equality. The takeover in effect was a battle between different and conflicting conceptions of community.

Similarly, Labaree's framework points to the way in which the takeover conflict was about more than community. One of the issues that divided Detroiters was their assessment of the success of the city's schools in educating their children. Those who supported the takeover pointed to the failure of the schools in providing for these children, while those opposing the takeover noted the improvements that were occurring in the efforts of the schools to educate these same children. In both instances, these counter claims were framed not in a communitarian discourse but in an individualistic one that viewed schooling as a private good that mediated the life chances of children in a competitive economy. The meaning, then, of conceptual lenses are themselves not easily settled. There are, as we saw in this chapter, multiple discourses that we can bring to bear to interpret school reform. And these discourses mean different things and lead us in different directions.[83]

VIII.

In this chapter, I examined the conflict surrounding the mayoral takeover of the Detroit Board of Education, looked at the role that race played in that dispute, and considered what the use of community as a conceptual lens tells us about this reform initiative. There was a major difference between the reforms that we considered in previous chapters in that there was an important shift in the role that race played. The conflict over More Effective Schools and the Clinic for Learning in New York City in chapter two and Detroit's Northern High School walkout in chapter three were conflicts that pitted urban Blacks against a White political and educational establishment.

The effort to establish a mayoral appointed board of education was different in that it was largely a Black-Black dispute. On one side stood a coalition that was largely composed of Blacks but included some Whites, particularly corporate groups, which supported the takeover as a means of restructuring the city's schools. And on the other side was a more decidedly but not exclusively Black group that opposed the takeover because they saw it as threatening the principles of local control and, at least for the African Americans in this coalition, as undermining their hard fought voting rights. The salience of race to this dispute becomes particularly clear when we recognize that much of the state's economic and political power that was in White hands supported the takeover.

Despite these clear ideological differences, it is important to note that much of the support and opposition to this reform was not ideological but rather came from individuals on both sides who defended their position on the pragmatic grounds of the efficacy of Detroit's schools in educating city children, particularly their own children. Labaree's description of the contradictory purposes that American schools often serve points to the heterogeneous makeup of both the supporters and opponents of the takeover.

In our discussion of the More Effective Schools Program and the Clinic for Learning in New York City during the 1960s, it was a Black-White conflict that undermined the growth of a sense of community. At the same time in Detroit, it was the discord between African Americans and the city's white educational leadership that promoted a spirit of community among that city's Blacks. The combination of the pragmatic quality of the debates surrounding Detroit's mayoral takeover controversy coupled with the fact that Blacks and Whites were on both sides of the conflict does seem to suggest that race may have played a very different role in this struggle. The nature of that role, however, is hard to discern. Opponents of the takeover labeled the proposal as a racist attack on Black Detroiters from Whites outside of the city, particularly Republicans. Detroit's Black House Democrats labeled Archer a traitor to Black interests for his support of the takeover. Even when opponents did not talk explicitly about race but instead criticized the takeover on the grounds that it violated local control and threatened the voting rights of Detroiters, race was on their mind. Local control in this context meant Black control and that along with the right to vote were seen by those opposed to the takeover as the fruits of the nation's civil rights revolution.

Our next chapter brings us to the present and to reform initiatives involving a new stakeholder. As I noted in chapter one, the social and economic transformations that are presently occurring in urban America under the rubric of globalization have changed the role of the state from that of governing to that of enabling. Under such conditions, a new player has entered the realm of school reform—the public private partnership. A creation of civil society as much as of the state, partnerships, as we shall see, are vehicles that are changing the terms in which discussion of school reform are taking place. Race has not disappeared from this terrain. Yet, it is an issue whose role while not diminished is less obvious.

Chapter 5

Educational Partnerships, Urban School Reform, and the Building of Community

At the beginning of April 2000, parents whose children attended five of the then worst performing schools in New York City voted to defeat an attempt to convert the schools to charter status. New York State's 1998 charter school legislation allowed for the conversion of existing public schools to charter schools with the approval of a majority of the parents of children attending the school.[1] Proposed by School Chancellor Howard Levy with the urgings of his boss, New York City's then Mayor Rudolph Giuliani, a supporter of privatization as a reform strategy, the plan would allow the private, for-profit Edison Schools to enter into a partnership with the city to direct the conversion process and, if successful, to become the educational management organization (EMO) responsible for administering these new charter schools.[2] The ultimate failure of this proposal, occurring in the midst of the greater success of another venture, the Annenberg Challenge, raises the question of the role that partnerships have come to play at the end of the twentieth century and the beginning of the twenty-first as a vehicle of urban school reform.

In this chapter, I will consider what we might learn about this approach to reform by looking at partnerships in New York City, Detroit, and Minneapolis through the lens offered by a concept of community. My starting point will be the idea of partnerships and their role in school reform. I will then turn to the struggle between Edison and the Association of Community Organizations for Reform Now (ACORN), a grass-roots community organization whose membership was comprised largely of low income families who worked to improve their economic and social conditions and to enhance their political power. I will next consider two Annenberg supported school reform projects, the New York Networks for School Renewal (NYNSR) and Detroit's Schools of the 21st Century. Finally, I will consider the work of an organization known as Youth Trust to develop and operate school-business partnerships in Minneapolis.

Partnerships refer to the attempts of individuals and groups from various sectors within both civil society and the state to join together to solve mutual problems. Such patterns of collaboration within American society have had a long history reaching back to at least the turn of the twentieth century and the efforts of charitable organizations of one sort or another, often working with government, to establish kindergartens, vocational schools, and schools for disabled children. Partnerships became a particularly popular vehicle during the 1960s to enlist the state and the voluntary sector in combating poverty and enhancing educational opportunity among minority and working class youth. Performance contracting in which school districts turned over the operation of one or more of their schools to for-profit businesses that would guarantee improved student performance as a condition of payment represented one form of such partnerships. A number of federal agencies that were part of the War on Poverty were another form of public private partnerships. The Job Corps, for example, involved such corporations as General Electric and Westinghouse working in cooperation with the U.S. Department of Labor in the management and operation of the agency's residential and non-residential vocational training centers. Still another form of partnership that was popular during this period were collaborations between school districts, universities and foundations in support of educational reform. I discussed two such ventures in chapter two. The Clinic for Learning involved a collaborative effort between New York University and the New York City Schools to enhance achievement at Brooklyn's Whitelaw Reid Junior High School. Another was the Ford Foundation's Gray Area Program that joined with a number of urban school districts including New York City in support of an array of educational reform initiatives.[3]

II.

In August of 2000, Chancellor Levy issued a request for proposals (RFP) from for-profit and not-for-profit organizations with experience in managing schools to oversee the conversion of five New York City schools to charter schools over the next year. As charter schools, they would operate under state authority and be supported by public funds but would possess far more autonomy than regular public schools. While these schools would be expected to adhere to the same standards regarding health and safety, civil rights, and assessment that governed all New York public schools, they would be freed from all other state and district regulations. Most importantly, they would have the autonomy to design their own educational program provided that it met state standards regarding student performance and it was consistent with the provisions of their charter.[4] The schools in question were two elementary schools, one in Harlem and one in the Bronx, two Brooklyn middle schools, and one intermediate school in Brooklyn. Each of these schools had been identified by the New York State Education Department as a School Under Registration Review (SURR), a designation for low performing schools that were required to improve or to be closed. Of the 105 schools that were on the SURR list at the beginning of 2000, 95 were New York City schools.[5]

The SURR list was only one indication of the problems facing New York's public schools.[6] In 1998, the city's schools had the fourth highest dropout rate among the nation's ten largest urban school districts. While 70 percent of its white students graduated high school in four years, only 42 percent of its Black students and 38 percent of its Latino/a students did so. Of all of its high school graduates, only 22 percent performed sufficiently high on the state competency tests to earn an academic diploma.[7] The vast majority of the city's high school graduates received a general diploma.[8]

Like other big city school systems, this pattern of low achievement was a product of a post–World War II demographic and economic transformation not dissimilar from the ones that I described for Detroit in chapter two. During the last fifty or so years, New York City, like its Midwestern counterpart, has experienced the demise of its blue collar industries and a resulting deterioration of its local economy, a significant population loss, the flight of a portion of its white middle class to the suburbs and their replacement by poor Black and Latino/a families, an increasing childhood poverty rate, an expansion of its welfare rolls, and a growth of single parent families. Moderating the effects of this transition in New York City was the fact that it was not a one industry town. The growth since the 1970s of its financial, knowledge, and information sectors have brought a resiliency to its economy that has benefited its middle class of all racial groups. Yet, for large numbers of minority New Yorkers, these changes have left them in economic distress and undermined their children's academic achievement.[9]

Under the terms of the chancellor's RFP, a successful bidder was required to work with the schools, parents, local organizations, and the surrounding communities to assess the desire of stakeholders for the conversion of their schools to charter status and to work with them to develop the required charter applications. The successful bidder was to receive compensation of $99,900 for each school for developing the conversion plan and securing parental approval and another $500,000 per school for the actual conversion.[10] There were 19 bidders and Edison was awarded the contract.

In late January of 2001, the board of education announced the selection of Edison as its conversion agent and set the beginning of March for parental voting. The next month, ACORN filed suit against the board and indicated its opposition to the conversion. The organization asked that the election be delayed to allow it time to mount its opposition campaign and that it be given resources equivalent to those given Edison to conduct this campaign. What followed were a series of negotiations between Chancellor Levy and ACORN that resulted in a consent agreement in which the organization was able to secure an extension of the voting date and some resources, although not equal to those provided to Edison, to promote its opposition.[11]

The dispute between ACORN on one side and Edison and the board of education on the other provides us a good picture of the kind of conflicts that can surround attempts to use partnerships to reform schools. For the ACORN official to whom I spoke, this was not simply the transformation of traditional schools into charter schools. It was an attempt to privatize these schools. Actually, it was not privatization so much that bothered ACORN. It was rather that this effort was being undertaken, in her words, on "the back of low income and minority neighborhoods" and, in fact, was being "rammed down people's throats." She pointed out that such policies were not being enacted in affluent, middle class areas of the city but only in poor and minority neighborhoods. She thought that the funds and office space that the board of education had provided Edison had given them an unfair advantage in promoting the conversion over those parents and community residents who were in opposition. She also charged Edison with conducting their campaign using unethical tactics. Edison, she claimed, had promised undocumented immigrant parents that if the conversion was successful, they would help them in dealing with the Immigration and Naturalization Service. Groups like Edison, she went on to say, were "poverty pimps" who were given tax breaks to change the schools in ways that would ultimately gentrify poor, minority communities.

As this ACORN official saw it, the effort to convert these schools to charter status was something of a conspiracy. She placed most of the blame on Mayor Giuliani who was, to her way of thinking, attempting to advance his own belief in privatization as a reform strategy. She claimed that the mayor actually wanted the chancellor to provide him with the names of one hundred low performing schools to privatize and then lowered the number to twenty. Levy, who she thought was not all that favorable toward the conversion, came up with his plan reluctantly as a way to assuage Giuliani. For the mayor it was, as she put it, "just another arrow in his quiver of privatization," and proof that he was not a friend of the poor.

This ACORN spokesperson did not, however, see Levy as blameless. He had created a special "Chancellor's District" for the SURR schools in the city whose purpose was to develop initiatives to improve their academic performance. And these efforts were actually working at the very moment that the conversion plan was being pushed forward. What troubled ACORN was that once Levy had placed the schools in this special category, he claimed that they were not improving fast enough and that the only viable option was to turn them over to Edison. There were, according to this ACORN informant, numerous high performing schools in New York City that, she believed, could have served as models for improving academic achievement in the schools in question. Fixing these schools, she went on to say, was not all that difficult. They required more resources, the presence of fully certified teachers, physical repairs, relief from overcrowding, and increased parent involvement. "Everybody in the city knows how to turn these schools around." What Levy was doing by bringing in Edison, in her opinion, was "beating up on parents."[12]

Two Edison vice presidents who I interviewed saw matters differently. As one of them noted, "it was a huge opportunity for us right in our backyard to demonstrate that we had an incredibly good product that could turn around the public schools and the community in a positive way." He went on to say that the reason that they were selected was because of the company's history and their ability to "invest huge sums of money." He disputed ACORN's claim that the conversion had anything to do with Mayor Giuliani. Although he was in charge of the conversion effort, this Edison executive claimed to never having spoken to the mayor about this issue. He also discounted the charge that Edison had acted unethically. There were among the parents in these five schools those with serious immigration problems. All the company did, he pointed out, was to offer those individuals access to lawyers who could advise them on such matters.

As these spokespersons for Edison saw it, the conversion dispute had three distinct causes. There were issues related to timing. The state's charter school legislation was only two years old and consequently virtually none of the key stakeholders had any experience with the creation of charter schools. They went on to say that the board of education did not keep the community informed about what was occurring in their search for a conversion agent, and as a consequence there was little in the way of information that was released between the issuance of the RFP and the announcement of Edison as the choice. A second issue involved the role of ACORN. From the vantage point of these spokespersons for Edison, ACORN was not a true stakeholder. The five schools involved in the conversion were all different and each was represented by different individuals and groups who espoused distinct positions. ACORN, they claimed, consciously chose to become involved and in fact was associated with a reform group, the New York Networks for School Reform, which saw itself as a competitor to Edison. ACORN, according to one of these Edison spokespersons, did not know much about the company and that in fact they had cancelled numerous meetings that had been set up between the two organizations.

As it turned out, there were ambiguities in the conversion proposal that exacerbated the disagreement between ACORN and Edison. At the heart of ACORN's opposition was their belief that ultimately Edison's task was to privatize these five low performing schools. Yet, it was unclear whether Edison's role was to direct the conversion or actually, if the campaign was successful, to manage these five newly created charter schools. The individuals to whom I talked from the Board of Education's Office of Charter Schools recognized this uncertainty. The chancellor's RFP required that the proposals submitted by bidders address both the conversion process to charter school status and the management of these new schools. Yet, it seemed to be left open as to whether the successful bidder would, in fact, be responsible for both tasks.[13]

Edison also complained that the United Federation of Teachers (UFT), the union representing city teachers, campaigned against the conversion. They accused the union of distributing anti-Edison flyers outside of two of the schools and inciting teachers against the plan. One of the UFT officials who I interviewed claimed to be reflecting the opinions of teachers who they represented and parents within the school in opposing the conversion. It was the absence of teacher and parent support that led the UFT to oppose the plan. Of particular importance to the union, according to one of these officials, was the fact that at least two of the schools that were slated for conversion had begun to turn around. They had, she noted, introduced new programs and had new principals. They just "needed a bit more time." She went on to say that the union "took the pulse of the teachers. Our opposition come from the staffs but…if the parents want to, you know, vote to go ahead with it, of course we'd support that."

III.

With the election approaching, both ACORN and Edison used paid canvassers to campaign in the neighborhoods surrounding the schools. ACORN took a more aggressive approach and charged that the conversion was a moneymaking scheme for Edison. As one ACORN organizer noted to parents in Brooklyn, "the thing about Edison is there's no heart there; it's all business and they will treat our children as products." Edison, on the other hand, took a low-keyed stance. At a meeting of supporters in Brooklyn, a company official pointed out that they did not want to engage ACORN in a debate. Their task, he told these canvassers, was to explain to parents the good work that Edison was doing in schools throughout the country. As he put it, "at the end of the day, parents want something that is better for their children." The campaign, however, was quite bitter. There were charges that ACORN organizers were spreading misinformation to the effect that if Edison was successful, it may charge tuition and it may expel students from their schools for poor performance. And there were counter charges that Edison was making their case by cornering parents in school yards and telephoning them late at night.[14]

Divisions were particularly apparent during the election itself, which involved ten days of telephone and internet voting beginning in mid-March and culminating with two days of in-person voting at the end of the month. On election day at one of the Harlem schools, two crowds of parents lined up outside the building's entrance. Those supporting the conversion shouted "vote for Edison," while those in opposition shouted back, "No, no Edison. Yes for education. We do not sell our children." The comments of two parents highlight this conflict. One of these parents, who had voted against Edison, commented that "I don't feel anybody should come in and make money off our children." Another parent voted in support of the conversion "because," as she put it, "I liked the way Edison plans to make smaller classrooms. I believe children learn more that way. Right now, my daughter has 36 students in her classroom."[15]

The campaign was an uphill battle for Edison. The company did have the support of the board of education, and Chancellor Levy sent a letter to affected parents urging them to approve the conversion. It also received editorial support from the *New York Times*.[16] ACORN countered by having two key Black New Yorkers, former Mayor David Dinkins and the Rev. Al Sharpton make recorded telephone calls to parents urging them to vote against the proposal.[17] According to the reports of Edison's canvassers, many of the parents they approached knew nothing about Edison. These parents were not satisfied with what was occurring in their children's schools. Yet, they were unfamiliar with Edison and did not know much about its program.[18] One of the Office of Charter Schools officials to whom I spoke commented that many of the parents did not even know that their child's school was having difficulties and likely to be closed. Especially unhelpful to Edison's cause was the decision by the San Francisco Board of Education in the midst of the election to revoke Edison's contract for operating a charter school in that city on the grounds that it had violated the terms of its agreement.[19]

Ultimately, ACORN had an easier time securing votes against the conversion than did Edison have success in securing support. The requirement in the state's charter legislation that conversions required support of more than fifty percent of the parents of children attending the school and not a majority of those voting did not help Edison because school board elections in New York City routinely get less than five percent voter turn out.[20] In the end, only 2,286 parents of the almost 5,000 eligible voters cast ballots with 1,833 voting against conversion and 453 voting in support.[21]

IV.

We can at this point in our account make an initial effort to look at partnerships from the vantage point of a notion of community. Of particular importance in this regard was the prospect that Edison could bring important stakeholders together around a common vision for the five schools in question. Clearly, Edison and ACORN had very different views about school reform. Writing in the *New York Times* a week after the election, Christopher Whittle, Edison's Chief Executive Officer, admitted his company's failure in conducting a campaign that demonstrated their commitment to improving the schools. "We were so excited about the opportunity to transform low-performing schools right in our own backyard that we agreed to a plan with flaws, however unintended by its authors." For Whittle, the conversion effort lacked community input and did not allow Edison to develop a true partnership with the board of education or with the parents of the five schools. "A far better approach," he commented, "would have been to allow schools to compete for selection. In communities where such a process has been used, parents, teachers, and neighborhoods had time to become acquainted with Edison's design and the resources we provide. The communities chosen were thoroughly involved, understood Edison and embraced our goals."[22] But Whittle's comments hardly cleared the air.

The ACORN official to whom I talked took exception to his altruistic claim. As she saw it, the only thing that Whittle was excited about was the money that Edison would make in converting these schools and then managing them as charters. Whittle's comments, in her words, were examples of "elitism" and "racism." A campaign for charter schools, at least ones that might be privatized, could not muster the kind of common understanding and purpose that is required to obtain the civic capacity necessary for the building of community.

A particular problem that seemed to face this partnership venture from the outset was the absence of any sense of common purpose among the various stakeholders. The ACORN official who I interviewed interpreted the defeat of the conversion plan as indicative of the ability of the struggle to mobilize poor and minority parents around a common agenda. "Nobody had a plan," she noted, "but throughout all of this noise, we have pulled together a consortium of folks who had guts, and that's the operative word, who said they would be willing to help the schools."[23] It is, however, not all that clear that such an agreement was in fact achieved. One of the interviewees from the Board of Education's Office of Charter Schools noted that the effort to convert five different schools had created five distinct and possibly unrelated disputes over conversion led by different constellations of opponents. In some cases, he noted that ACORN was the key player, but in others it was community members having no affiliation with this organization. He went on to say that while ACORN saw Edison as its enemy, this was not the case for all the opponents. Some of these groups were supportive of charter schools but did not support this particular plan. Other groups were not all that opposed to Edison but did not like the idea of turning their schools into for-profit entities. In short, he saw the conversion campaign as a much more complex and complicated struggle involving more than just ACORN. In the end, ACORN was no more able to mobilize the level of civic capacity required for community than was Edison.

V.

Our treatment of partnerships thus far leaves us with the question of whether partnerships can in fact be vehicles for building community. To explore this issue, I will now consider a less contentious effort to build a partnership, the Annenberg Challenge.[24] Launched in 1993 by publisher, philanthropist, and former ambassador to Great Britain Walter Annenberg, the Challenge involved a gift of $500 million in support of eighteen locally developed projects to improve urban and rural education as well as to enhance the role of the arts as a critical element in basic education. Among the awards were grants in amounts of $10 million to $53 million to support reforms in a number of urban school districts, including New York, Boston, Chicago, Houston, Los Angeles, the San Francisco Bay Area, and South Florida, and through its Rural Challenge to numerous communities throughout the nation; awards in amounts between $1 million to $4 million to several smaller urban school systems, including Atlanta, Baltimore, Chattanooga, Salt Lake City, and Chelsea, Massachusetts; and a number of awards to support education in the arts in Minneapolis, New York City, and several regional consortia throughout the nation.[25]

For its proponents, a distinguishing feature of the Challenge, which they claimed sets it apart from reform efforts involving both centralization and privatization, was its reliance on so-called intermediary organizations as its driving force. Such organizations occupy something of a middle ground between two agencies, in this instance, school districts on the one hand and individual schools on the other and from that position direct changes in both organizations. Intermediary agencies serve among other things to promote school reform policies, to mobilize and to bring together the diverse constituencies that support school reform, to assist in raising public and private funds for such efforts, and to define accountability standards for reform. These are agencies that ostensibly serve to enhance the capacity of reform initiatives to fulfill their missions.[26] The Annenberg Challenge had great faith in such organizations. In the school reform projects that they supported, funding did not go directly to school districts or schools but rather to these intermediaries who, in turn, directed it toward specific school and school district projects.[27] In the remainder of this chapter, I will employ the lens offered by the notion of community to explore intermediary organizations as a kind of partnership venture to bring about successful school reform.

The intermediary organization that directed the Challenge in New York City was the New York Networks for School Renewal (NYNSR). A collaborative of three intermediaries and a community organization, NYNSR offers both a picture of the involvement of different groups with distinct reform agendas and a site where reform did get intermingled with urban school politics. Under the terms of the grant, the Networks received $25 million over the course of five years that was later extended for another year and a half. They were required to supplement this grant with a two to one match, half coming from private funds that these organizations themselves raised and half coming from the New York City Public Schools.[28]

The roots of this project go back to 1974 in East Harlem where Deborah Meier and a group of other progressive minded educators established

Central Park East, an alternative elementary school of around 150 children reflecting the principles of child-centered education and a Deweyan philosophy. The school was built on the premise that a small institution offering its students a themed curriculum and personalized attention and providing its staff with a large degree of autonomy from district rules and procedures could enhance academic achievement. The next ten years saw the extension of this concept to other schools, and in 1987 six of these small schools throughout the New York City joined together into a network, the Center for Collaborative Education (CCE), to advocate for school change as well as to provide technical assistance and support for schools undergoing such transformations.[29] These schools, according to one of the founders of the Center shared "a certain set of principles and beliefs about teaching and learning." Their coming together, she noted, represented an attempt, in her words, to "support each other and also help get the word out what we are doing."[30] The next year saw the transformation of what was the fund-raising arm for the chancellor of the city schools, the Foundation for New York City Public Education, into an independent organization supporting small schools, New Visions for Public Schools. Once an agency that served to channel money into the public schools, the organization became an intermediary that linked the public schools with corporate and philanthropic agencies committed to school reform.[31] And in 1989, still another intermediary committed to the idea of small schools, the Center for Educational Innovation (CEI), was created. Eleven years later CEI would merge with yet another intermediary organization, the Public Education Association, which had been supporting school reform in New York City since 1895, creating CEI-PEA.[32]

In 1993, these three organizations along with the New York chapter of the ACORN came together under the banner of NYNSR to launch the city's Annenberg Challenge project.[33] Unlike the other members of the Networks, ACORN was not a typical intermediary but rather more akin to a political organization with a civil right agenda. The impetus behind this initiative was the commitment of the four partners within NYNSR to a set of common beliefs about quality urban education. They supported public school choice. They favored the establishment of small schools that offered a curriculum organized around themes that were appealing to children, that engaged students in active learning, and that held students to high standards of achievement. The Networks believed that teachers required sufficient autonomy to be able to make decisions about teaching and learning that met the specific needs of their students. They believed that schools should provide students with personalized attention. They supported parent and community involvement in the education of children. And the organization argued that teachers should be provided with the resources and professional development opportunities required to ensure high student achievement.[34]

VI.

The major achievement of the project during its five years of existence was the establishment of almost 140 small schools throughout New York City that served approximately 50,000 students.[35] The majority of these schools ranged in size from about 250 to 500 students, although there were schools within the Networks that were as small as 50 and as large as 2,000 students. The schools were affiliated with one of the four intermediary organizations, joined together around common characteristics, and often embraced a specific theme or educational philosophy. There were, for examples, schools whose curriculum and teaching practices reflected a commitment to, among other things, social justice, to the arts, to community involvement, and to ecological awareness. Schools within the Networks ranged in orientation from progressive to traditional and to any of a number of viewpoints in-between, emphasized college preparatory as well as career oriented programs, and served a range of students from at-risk to gifted and talented.[36] At least initially, according to two of the intermediary group officials who were interviewed, their organizational plans were more ambitious and called for the development of what they called learning zones or clusters of schools in geographical proximity to each other, but those plans never materialized.

One such school is Bread and Roses Integrated Arts High School, located in central Harlem and affiliated with New Visions and ACORN. Named after the ballad that celebrates the 1912 textile workers strike in Lawrence, Massachusetts, the school, with an enrollment of about 350 students in grades nine to twelve, seeks to integrate issues of social justice with a focus on the creative arts. Although the curriculum was academically oriented and college preparatory, the school includes both learning disabled students and second language learners in heterogeneously grouped classes.[37] The principal noted that in response to the 2001 destruction of the World Trade Center, students were engaged in a project in which they constructed murals that depicted the "...island of Manhattan and all the work that goes on over it and underneath it...So there's definitely an emphasis on supporting and seeing themselves as part of the working class, as part of a class that's very powerful and can make this a just society for everyone." As this project suggests, artistic activity, specifically mural construction, was being used to develop an understanding of social justice, in this case the role of the working classes in American society. As this principal saw it, this was a different approach than is typically available in other city schools that provide students with both academic courses and classes in art. "What we do that's really unique," she went on to say, "is that we encourage collaboration between artists and academic teachers. So kids have an opportunity to not have just sort of chalk and talk experiences in classrooms but instead get an opportunity to actually utilize a creative expression to show what they've learned. So their exhibitions often include posters or a play or poetry or a song they've written."

Also located in central Harlem is Frederick Douglass Academy, which serves around 1,000 students in grades six to twelve. Affiliated with the Center for Educational Innovation (CEI-PEA), the school features a rigorous college preparatory curriculum that includes classes in Japanese and Latin in the middle grades and numerous advanced placement courses and introductory college courses in the high school. Founded on the belief that all children, notwithstanding their economic and family circumstances, can achieve academically, the school requires students to adhere to a student creed that promotes excellence. The creed notes that Frederick Douglass is "a community of scholars" in which its members are obligated to adhere to "a code of civilized behavior" and "to refrain from and discourage behaviors which threaten the freedom and respect every individual deserves."[38] In addition, students are required to subscribe to twelve "non-negotiable rules and regulations" that stress responsibility and good behavior, wear a school uniform, and complete four hours of homework every night. The overriding commitment of the school is to prepare what the head of CEI-PEA describes as the "full range of students" from Harlem for admission to college. "This is," he points out, "not an elite school, but it gets elite results." In 2001, Frederick Douglas graduated 114 students of which 113 entered college.[39]

Beyond its success in establishing small schools, NYNSR points to a number of other important achievements. The Challenge's final evaluation, which included analyses of outcomes in comparison to other city schools, surveys, interviews, and participant observation, indicates that during the course of the Annenberg grant, student academic achievement as well as standardized test scores improved and frequently surpassed citywide averages. There was some up and down fluctuation in performance during these five years, but at its conclusion a greater proportion of the Networks' students than at the beginning were achieving at or above citywide averages. White students enrolled in the Networks' schools did outperform Black, Asian, and Latino/a students. Yet, there were consistent improvements in Black and Latino/a achievement throughout the course of the project. Graduation rates at NYNSR high schools over the course of the grant were at least equal and sometimes better than the graduation rates at other city high schools. Similarly, these schools have experienced significantly lower dropout rates than other city high schools.

Schools within the Networks reported a number of positive structural changes. They noted that they were able to introduce a rigorous and coherent program that was innovative, particularly with respect to the inclusion of experiential learning. They pointed to their ability to create student-centered schools that featured learning environments that promoted teacher-student collaboration. They also noted their success in establishing a culture of shared leadership between administrators and teachers.[40]

VII.

Annenberg money, which for individual schools amounted to around $10,000 to $25,000 per year, enabled the participants to do things that they had not heretofore been able to do. One of the most important of these initiatives was the provision of funds to support teacher professional development. The principal of one of the participating high schools noted that her location in Central Harlem was not a place where people wanted to teach. She felt that her teachers, many of whom were young, "are passionate and committed to teaching, but they really have no experience in the classroom." The funding enabled the school to bring in experienced teachers who could provide staff development. In this vein, the principal pointed out that "as a brand new school, it allowed us to focus on curriculum and instruction as opposed to trying to scavenge for money."

"What we've done using Annenberg money," she went on to say, "is to be able to use folks in the community who are artists...who we wanted out kids to be exposed to." The money enabled the school to bring in musicians, actors, and others who could work to "build the arts component of our school." There were, she pointed out, numerous performing artists in Harlem who could assist the school, but the board of education did not have any money to support such an effort. Annenberg funds also enabled the school to construct an art gallery in the library. "Kids," she commented, "have a real respect for art. They don't have to go downtown in order to see real art. They can see it in our library." The principal also noted that Annenberg money enabled the school to rent space for its first graduation that was sufficiently large to allow "us to bring every kid in the school and every parent to see our kids walk down that aisle." She went on to say that doing so was really important to a new school like hers "that has no track record." For an elementary principal participating in the Challenge, the support enabled him to hold retreats for his staff, to send his teachers to conferences, and to establish a parent library.

In addition to Annenberg, participating schools often had other partnerships. The principal of one of the participating high schools mentioned that his school had established a partnership with the Goldman Sachs investment management firm that provided funds to allow graduating seniors at the school to develop portfolios to showcase their academic work. Partnerships were not, however, always that popular with the participating schools. One of the elementary school principals in the project commented that "I'm not interested in partnerships because I don't like the whole funding game. I don't like to have to go through hoops." He understood that a potential partner organization would want a school to embark on an initiative of interest to them in return for its money. He felt, however, that such an organization should "give us the money and let us create the programs. Don't make me go through [something]... that fits your specifications." He went on to say that "I want someone who's just going to support us, you know, spiritually, intellectually, financially, organizationally, without any strings. I'd appreciate someone who would challenge us on all of those levels. I'd like someone to say, here's $100,000 a year you can have to do things in your school... I don't want someone to say here's $100,000; you have to do X, Y, and Z."

Another elementary principal expressed a similar view. As he saw it, reform has to come from within by those who recognize the needs of the schools. He went on to say, however, that "schools can often just get into carping and blame unless they have some help from without." Successful reform, he believed, requires both an internal impetus and external assistance. The problem, however, was the amount of time and work that was necessary to create a partnership. "It's a massive headache, and you never get the time... Yeah, I think it's self-defeating in a way. There has to be a way, it seems to me, if you are going to do partnerships, if you're going to rely on partnerships as a vehicle, a mechanism, that you free people in administrative positions to be able to dream a little bit, to think about things." Ultimately, for this administrator, partnerships represented a mixed initiative. On the positive side, developing a collaborative relationship with an external partner required teachers and administrators to look at the school through "other people's eyes." On the downside, however, partnerships can lead the school away from its stated commitments to embrace the views of a potential supporter.

A major concern of these participating principles was the ending of the grant. As one of the high school principals put it, this was "disastrous because not only have the Annenberg funds been used up...but the New York City budget has been cut by millions of dollars." For her school, it meant that many of the services that had been provided with Annenberg support would no longer exist. "We can't hire any consultants to work with our teachers. Our staff development has to be pruned back so that there aren't consultants who come in." She noted in that vein that without Annenberg money, she was no longer able to fund the consultant who had come in to conduct workshops on writing across the curriculum. It was not, however, only staff development that suffered. The ending of the project "impacts on everything."

An elementary principal described the demise of the project as "being kicked out of the house and having to stand on your own feet again... This year, you know, is the first year I can really feel it. Last year, I didn't think we would get the money and then there was some money that materialized that made it easier for me to get things done. So it made it easier for us to have a story telling component within the school. Whatever our plans were, it helped me to do things." The ending of the project has meant that he has to rethink what he is doing so that "we can pay for everything with the money that we have." Making matters more difficult, he noted, has been state and city budget problems and the World Trade Center disaster, which taken together have reduced available money even more. According to a principal of another participating elementary school, with the demise of the Annenberg Challenge, "we don't have a penny to pay for some very important stuff." The absence of money will not, he went on to say, end all renewal efforts, but it will make them more difficult.

Not all the project participants, however, felt this way. The principal of one of the project's high schools saw the Annenberg support as seed money to enable her school to launch its innovations. She fully recognized that once the funding was gone, she would have to do additional fund raising. This was, as she saw it, part of her job. Another high school principal held a similar viewpoint. In order to be successful, he noted, he had to be something of a "hustler." He was proud of the fact that he was able to convince the city council to provide his school with $900,000 to equip a science room. He was, however, even more proud of the fact that through cutting corners, he was able to use this money to equip four such rooms. From his vantage point, "if someone says no, you know that you gotta go around them. If they say no, there is a yes somewhere." The principal of a participating high school felt, however, that the availability of Annenberg money made it unnecessary for him to deal with the board of education's bureaucracy. He did not have to get three bids every time he wanted to make a purchase. He simply went to the appropriate store and bought what he needed.

VIII.

The concept of community does provide us with a lens for interpreting what was occurring under the auspices of the New York Networks. The role of the intermediaries within this project was to construct a unified community that would activate sufficient civic capacity to propel forward school reform. There are indications that at least within the Networks such capacity was evident. The Networks were, as I noted earlier, comprised of four separate and distinct intermediary organizations with different agendas. A New Visions administrator commented that "we work very closely with the board of education, which makes us different from the other agencies because we do not air our differences publicly. We seek to find consensus with the board of education on very tricky issues—to work behind the scenes as opposed to publicly challenging them. We see them as an ally. This does not mean that we don't have differences. We just chose to handle them differently than other organizations."

In contrast, according to an official from NYNSR, "CCE works outside that box in the sense that they work with individual superintendents and are much more top down in the work they do but don't necessarily want to collaborate with the central board as much as New Visions does. [And] ACORN is a grassroots community organization that is very parent-focused." A founding member of CEI-PEA noted key differences between his organization and other participants in the Networks. As he saw it, New Visions was the fund raising arm of the school chancellor, and CCE's uniqueness lay in its commitment to a particular philosophy. His organization, on the other hand, did not have a single orientation. "There are," he commented, "many ways to learn, and we want diverse schools... We are action oriented and interested in outcomes." In that vein, he noted that CEI sponsors schools that reflect Debbie Meier's commitment to child-centered education as well as schools that require a student dress code and strict disciplinary standards.

Despite these differences, these organizations were able to work reasonably well together. A staff member of the Networks noted that it was "not always smooth sailing, but they have gotten a lot done." In part this was due to the fact, according to the director of one of the intermediaries, that all of the key players in the organization knew each other and had worked together on educational projects for a long time. They were, he went on to say, not competitors, but were able to find common ground. "There is so much to be done. You never have to worry in the school system about the competition. The sad thing is that there's so few organizations in the country who are successful in the inner city. There's like a handful who really can say I changed the lives of these kids." Consequently, as he saw it, there was room for several groups approaching the problem in different ways. The strength of the Networks, in his opinion, was its diversity. Each of the partners were free, he noted, "to go about it the way you want to go about it...But in the final analysis you got to demonstrate that our kids are learning much better than the kids in surrounding communities. If the four organizations had a common mantra, it was, he stated, "greater autonomy for meaningful accountability as measured by pupil achievement and performance."

And, according to the director of another intermediary, all of the organizations held "a belief that schools could only be transformed if there was broad community participation in reform. We felt that the climate was right for community organizations to play a leading role in the reform of the New York City Schools. Philosophically, we believe that this project would help us to achieve our goals, which is to involve the parents and children in the educational system." He went on to say that the establishment of a network made sense because at the time there were a number of organizations that were working on educational reform in the city. "All of the organizations," he noted, brought "strength and brought different, diverse perceptions." As the director of a third intermediary saw it, there were a lot of schools in New York City and no one organization could deal with them. It required a partnership.

The fact that all of these organizations were joined together into a single, overarching coordinating body, the Networks, enhanced the ability of the groups to work together. As a director of one of the intermediaries pointed out, NYNSR was seen as a single entity, not a group of separate organizations. A Networks' staff member pointed out that the bringing together of the four intermediaries, the public school bureaucracy, the many participating schools, and an array of voluntary and corporate partners created a situation where "everyone speaks a different language." What was necessary, she went on to say, was the presence of a translator to ensure that the views of each participant were adequately communicated to other participants. And this, she noted, was the key contribution of the staff of the Networks to the project. What could have been a particularly divisive problem, the distribution of grant monies, was prevented at the outset by the appointment of one of the intermediaries, New Visions, as the project's single fiscal agent and an agreement at the beginning of how the funds would be divided between the four participating organizations.

If the existence of civic capacity defined the relationship between the intermediary organizations participating in the Networks, the same cannot be said for the relationship between the Networks and the New York City Board of Education. Administrators and teachers in the Networks' schools noted their limited authority in making important educational decisions. They were able to make decisions in certain areas, hiring of new teachers, establishing disciplinary policy, planning professional development, and evaluating teachers. There were, however, areas in which they had little influence, including determining curriculum, student assessment, and graduation policies. These remained issues under the purview of authorities outside the participating schools, namely the community district office, the citywide board of education, and the state department of education. A majority of administrators and teachers in project schools reported that there was conflict with different layers of the city's and state's educational bureaucracy. They did not believe that the city or the state did much in the way of supporting their authority or assisting them in their leadership responsibilities.[41]

The working relationships between Challenge administrators and the board of education were never all that good. The principal of one of the project high schools described the board as a "faceless bureaucracy." She went on to say that you can call the board, "but you have no clue as to who you are speaking with. It feels like you're sort of out here in the community, and they're over there. And it's not clear other than processing paperwork why they're there. It doesn't feel like they really want to support you." Working with the intermediary with which the school was affiliated, she noted, was far different. "With New Visions, it was clear from the very beginning that they wanted us to succeed. And they came through for us. Every time we had an issue, they came through for us. They were always on our side, even if it meant being on our side against the bureaucracy. They were on our side. It was clear that we were part of a team. They really saw it as being in their interest to help us. And that's what you need. If you're in a school, you need to know that there are organizations out there that really support you and aren't just finding ways to get you, you know what I mean."

This principal went on to say that she had never gone to a meeting at the board of education where she felt "respected." This was not the situation at New Vision where her and her colleagues "were treated like people who had something to say, and that's because New Visions came here and saw the work that we were doing." She saw the relationship with her intermediary as a professional, caring, and ethical one. With the board of education, on the other hand, it was one of, as she put it, "let's duck...and keep out of the way of whatever missile they're sending this way."

Administrators who led schools participating in the Challenge were also critical of the community district offices to which they reported. A 1998 survey of New Vision affiliated schools indicated that their principals resented the fact that they were often treated by their district administrators exactly the same as the principals of the larger, traditional schools within the district. Even though principals of these small schools often did not have the administrative help available in a large school with several assistant principals and other staff, they were expected to deal with the same amount of paperwork, file the same reports, and attend the myriad of meetings as other school leaders in the district. These New Vision affiliated principals did not feel that the issues affecting schools of several thousand students had anything to do with them and their schools. At the same time, district level administrators reported their displeasure with having to work with these small schools. They often were not very sympathetic to the idea behind creating small schools and disliked having to adjust practices and procedures that worked for most schools within their district to the unique needs of these small schools.[42]

We can also look at what was going on within the participating schools from the perspective provided by a concept of community. At Bread and Roses, for example, students were explicitly taught that they were members of a community that shared common values and a common purpose. In one of the school's chemistry classes last year, the principal noted that the students were introduced to issues of environmental justice. One of the class projects involved the students in a study of the high rates of asthma among children in the section of Manhattan where the school was located, which was attributed to the presence in the vicinity of the school of a city bus garage and a waste disposal facility. "So the kids," the principal went on to say, "looked at the correlation between asthma and what the science teacher felt was environmental racism. So that's an opportunity to survey our population and find out how many of our kids actually suffered from asthma and also go into the community and find out how prevalent that condition was in the community." As a description of the school that is distributed to prospective students and parents notes, "education is always a political act. It can either give people the capacity to transform their lives and the world around them or teach them to be passive recipients of knowledge that simply accepts the world as it is...Students need a curriculum and school committed to empowering them as learners and as constructive responsible citizens."

Located in the same building, another small school, Thurgood Marshall, pursued a similar community building approach. Marshall's curriculum, according to its principal was organized around the theme of social justice. Students pursued projects that explored topics affecting their community. Like their counterparts at Bread and Roses, they undertook a project to examine the high rate of asthma in Central Harlem. They also conducted surveys to explore how racist practices affected the neighborhood surrounding the school. As the principal saw it, the efforts of the school to enhance academic achievement did not have to result in students leaving Harlem for more affluent areas of the city. The school could in fact teach its students to empower their community. The pedagogy promoted in these two schools, then, sought to join students with the members of their community in a common purpose. From the vantage point offered by a notion of community, it would seem that the record of the Challenge in mobilizing civic capacity was mixed. There were changes in pedagogy and instruction that suggested that schools within the Networks were successful in attaining a degree of civic capacity associated with the notion of community. Yet, that kind of common purpose was never realized in the relationship of Challenge schools to the New York City Board of Education.

IX.

Half a continent away in Detroit another Annenberg Challenge project, Detroit's Schools of the 21st Century, provides another example of how partnerships are being used as vehicles for school reform. The Detroit initiative, which was launched in 1997 and concluded in 2004, involved a $20 million grant from Annenberg that required, as did the other Annenberg projects, a two to one match or another $40 million that Detroit would raise from public and private sources. The driving force behind the Detroit Challenge, much like that in New York, was an intermediary organization, the Schools of the 21st Century Corporation. Standing between the city's public school administration and the participating Challenge schools, the Corporation included a board of directors that was responsible for administering grants, raising additional money, and establishing policies whose members included the major public and private stakeholders for the initiative and a council, which was also comprised of stakeholders and served as the program planning and monitoring unit for the project. The project also included a Technical Support Consortium housed at Wayne State University to provide participating schools with assistance and training in implementing their grants and an External Evaluation Team from Western Michigan University to periodically assess the project.[43]

Not unlike the New York Challenge, the Detroit project was an effort to bring together an array of partners throughout the city to support school reform. The Challenge's associate director thought that the Detroit Challenge was unique in that it brought in as partners "all of the key players in the city—United Way, NAACP, the CEO of the Detroit Public

Schools, the teachers union, and the Urban League." The participants also included several major foundations, including the Kresge, Skillman, and Kellogg foundations, that provided a good portion of the required matching funds. There were a few corporate partners including American Telephone and Telegraph and DHT Transportation. The project, again according to the associate director, did not actively seek corporate support. He noted that a decade earlier, the city's corporate community had provided several million dollars to fund the Detroit Compact, a joint venture between business and the city's schools. "Frankly, the results of the Compact were not what we had all hoped," and consequently the city's business community "wasn't thrilled about stepping up and investing millions of dollars" in another educational reform venture.[44]

The schools that participated in the challenge also had their own partnerships. Within the ten clusters of schools that made up the Detroit Challenge, there were over a hundred partners. Most of the partners were community organizations that provided services that were directly related to enhancing academic achievement including tutoring, computer training, and after-school programs. There were also partners that provided the schools with medical, dental, and mental health services, and businesses and educational organizations that brought volunteers into schools to serve as tutors and mentors.[45] One participating school reported that it had established twenty-one such partnerships including links with Oakland University, Ford Motor Company, WJLB Radio, Lenscrafters, and RAPCO Substance Abuse Treatment Center. It was these partnerships that, among other things, enabled the school to provide volunteer tutors, to host school parties, to conduct workshops for teachers and parents, and to provide medical provisions for needy students.[46]

The initiative was launched in 1997 with a citywide competition for small planning grants to enable clusters of three to five schools to develop collaborative proposals for reform. The clusters were comprised of schools that had worked together on earlier initiatives or were new clusters that shared a common vision of school reform. These proposals were designed to address three broad goals that included improving teaching and learning; establishing better relationships among schools, parents, and communities; and creating more flexible governance practices between individual schools and central administration. Over 90 percent of the city's schools involving sixty-two clusters submitted proposals. From this first round of proposals, twenty clusters were selected to receive larger planning grants to further refine their reform initiatives and from these, ten clusters comprising forty-two especially designated Leadership Schools were awarded implementation grants that ranged from $2.1 to 2.6 million.[47] As the principal of one participating school pointed out, joining the Challenge required three grants. "The first grant got you the money to organize the people to write the second grant. That gave you the money to get the parents involved, to get everyone involved, to get the community involved... [and] to educate everybody to write the big grant."

In developing their grants, the successful clusters formed coordinating committees comprised of administrators, teachers, students, and community partners to work together to implement their proposal. Their key task in this regard was to select a research-based model of whole school reform to direct their efforts. Some, including High Schools that Work, Microsociety, and the Comer School Development Program, were organizational models whose purpose was to restructure the schools. Others, such as Direct Instruction, Success for All, and Different Ways of Knowing, were curriculum models designed to address issues of academic achievement. As the project's associate director saw it, these were models "that have been tried in other parts of the country and are shown to improve student achievement." He went on to say that "the driving force was that we wanted this to be a data driven initiative, and we wanted to impart the philosophy to the teachers, parents, and staffs of these schools that the effective use of data improves instruction." The models, then, were, in his words, a "kind of catalyst in getting the school community to buy into something." The clusters themselves took different forms, some horizontal comprised of a high school and its feeder elementary and middle schools and others vertical that joined schools with the same constellation of grades across feeder boundaries.[48]

X.

Cluster # 57 in Southwest Detroit adopted the Comer School Development Program, a model created by James Comer at Yale University that sought to bring together schools, parents, and the community to promote the social, physical, cognitive, ethical, and language development of children and youth. As Comer saw it, this was an approach that sought to provide the same quality of childhood development and reach the same level of achievement that he believed was once provided in small towns through family and community social networks. It was in effect a model in which schools were to assume the responsibility for providing the social support that in an earlier day and age families and close-knit communities provided to children.[49] The Cluster, for example, used some of its funding to establish a community health center and a center for community education.[50] This approach nicely matched an important goal of the Detroit project, to reestablish an older entity, "school-anchored neighborhoods." It was the desire of its proponents to establish such neighborhoods that led them to organize the project into clusters.[51]

There were three schools within the cluster, Logan and Beard schools serving grades pre-kindergarten to grade five, and Academy of the Americas enrolling students in grades pre-kindergarten to grade eight. It was, according to Beard's teacher facilitator, the principal of Beard who had first heard and read about the program and convinced the principals of the other two schools to join together. She noted that they then had to secure at least an eighty percent favorable vote for the model from each school's faculty before embarking on the program. As it turned out, there was little opposition. According to the facilitator, "I had some teachers that were very vocal about not wanting this, and several of them have retired."

The majority of students enrolled in each of these schools was Latino/a with over 80 percent of the students qualifying for free and reduced lunch. Operating within the parameters of the Comer model, the three schools comprising the cluster selected different curricular and pedagogical strategies. Logan employed interactive technology to develop learning centers to improve writing instruction. Academy of the Americas directed its efforts toward aligning its curriculum, and Beard employed a computer-based reading program, Accelerated Reading, to improve student reading skills.[52]

Beard's teacher facilitator estimated that during the first three and a half years of the project, the Challenge brought about $2.4 million into the three schools. For Beard alone, she went on to say, the money enabled the school to hire two additional reading recovery teachers, a bilingual teacher, and several teacher aides. It enabled the school to equip a new computer lab, purchase a scanner, to acquire the computer based Accelerated Reading Program, to fund a summer program, and to provide parent education opportunities.

The project also supported a range of professional development opportunities for teachers. The teacher facilitator at Logan noted that "we've sent four or five people for model training this year, which was $1,000. That's $5,000 just in registration fees. It's just, it's really a huge chunk of money devoted to professional development and training, which never would've occurred otherwise. And they come back into the building all excited and generate enthusiasm among the ranks to do, to do things that they might've considered before not to be so..." The training, she went on to say gave these staff members a degree of ownership in the program, which led them to take on responsibilities outside their regular duties.

The Challenge, however, brought in more than money. It enabled the staffs of the three schools within the cluster to look beyond their own setting and work collaboratively with the staffs of the other schools. The Beard teacher facilitator commented that what "21st Century really wanted was not just one school applying for this grant. They wanted you to try to work as a partnership and to do planning together and to try to share resources. And we've really done a lot of that. Like when we had our last workshop, we had three teachers from all three schools who were presenting. And then there were the three teachers we had sent for [Comer] training, and then they broke up into groups. We trained a first grade teacher; Logan trained a fifth; and Academy [trained] a third grade teacher. So you have a teacher from another staff that's, you know, been trained to help your teachers."

The teacher facilitator at Logan also saw benefits to the program beyond the money. "I think that it's not the Annenberg money so much. It's the reform model. I think that's really had more of an impact on the school... One of the goals of the grant was to improve self governance, of how we governed ourselves. And the reform model provided a structure to do that, that was not here before. Our meetings were haphazard. They weren't regular. Notes were not taken and distributed and that's completely changed now. An agenda is done before hand. It has to be approved; it's distributed. A meeting takes place that's very specific as to when it begins and ends and how it functions. And then notes are taken and distributed, and a log is kept."

For the staff in Cluster # 57, the Challenge had clearly brought new and needed resources into their schools. The question for them was whether it was worth the effort. The principal of one of the cluster schools noted that writing the initial grant was quite time consuming. She estimated that during the period the three schools were writing various elements of their grant proposal, she attended over seventy meetings. A teacher facilitator at another school commented that while the grant was being written, she was still a classroom teacher. She had a mother in a nursing home at the time, and it was, in her words, "overwhelming." "I mean, it really was two years of my life. We were working on this until seven or eight, two or three nights a week." As this teacher facilitator saw it, it took a long time to get the project up and running. She noted that "it takes three years to even get to the first stage of change. And we're now getting to the point where I see that everything is coming together. That we're working as a leadership team with all three schools. We go to meetings and it's like our meetings take twenty minutes... where they used to take two to two and half hours because the basic ground work has been done and we can just go on."

This teacher facilitator commented that "if the state or federal government treated all children the same and gave us what Northville and Novi and West Bloomfield have, maybe the need for partnerships wouldn't be as great." Since this was not, however, the case, she felt that Detroit schools needed partners if they were able to meet the needs of their students. What worried her, however, was the limited duration of the Challenge and what would happen when the funding ceased. Another teacher facilitator had the same fear. Without the grant, she noted, her school would lose their bilingual aides who assist in the office. She also pointed out that the Challenge had supported her position and that if she returned to the classroom, she could not undertake the array of responsibilities that she had taken on as a teacher facilitator.

There were, according to Beard's teacher facilitator, elements of the program that would be sustainable without Annenberg money. The computers that the schools purchased with Annenberg funds would remain. The ending of the grant, however, meant that there are "things that will not be able to happen. Like we hired a lot of parents... One of our initiatives was to get a lot of help to the first grade because we knew that's where the help [was needed]. At Beard, we hired three aides to work with the first grade for three hours. And it only cost us $5.75 an hour. So, it was a minimal amount of money for a lot of help for the first grade children." This was an expenditure, she pointed out, that would not continue after the expiration of the grant but had, she claimed, cut the first grade retention rate in half. Other initiatives that would be negatively affected when the grant ended, unless the school was able to find other funding, included a host of professional development programs as well as salaries for additional teachers, aides, and even the teacher facilitator herself.

In this vein, the principal at one of the cluster schools was particularly upset that the ending of the grant would mean that she could no longer afford to pay her teacher facilitator. "I'm going to tell you, we aren't gonna have the money to buy [her]. And when [she] stops, a whole mess of stuff stops... I'll tell you what I've seen. I've seen mothers who used to sit there and stare into space and gossip. Now they are feeling more empowered because they're paid a little extra; they're able to feel part of this whole thing, and I've seen real major changes."

The Challenge, she went on to say, was unrealistic in what they expected of participating schools. At the outset, she pointed out, the funding for the grant came in October although school had begun the previous month. In effect, the schools could not hire the personnel to support aspects of the grant until the academic year was underway. The school's teacher facilitator noted in the same vein that the grant will end in February of 2004 without any plans in place to pay those who were supposed to be supported by the grant until the end of the year. Similarly, she commented, the Challenge wanted the schools to sustain their efforts after grant funding comes to an end but offered no concrete ideas save to advise the school to secure community or business partners who would continue the funding.

A critical question in assessing the outcome of the Challenge was its impact on student achievement. In its final report, project officials noted that the participating schools did not reach the levels of student achievement that they had expected at the outset. From their vantage point at the end of the project, it now seemed unrealistic to them to expect major gains in student achievement during the relatively short period between the time the effort was up and running and its culmination. As they saw it, "the Leadership Schools were just at the beginning of the journey when project funding ended." By the end of the project, standardized tests scores in participating schools were increasing, but the results were, in the words of the final report, "spotty."[53] In Cluster # 57, for example, state competency test scores at the beginning of the project in fourth grade reading and mathematics at Logan and Beard schools exceeded district and state averages while those at Academy of the Americas were below those averages. At the end of the project, however, only Logan's scores remained above district and state averages. The scores at Academy of the Americas increased during the course of the project but remained below district and state averages. Toward the end of the project, Beard was transformed from an elementary school to an early childhood center and no scores were reported in 2004. The school's 2003 fourth grade reading and mathematics state competency test scores were, however, below district and state averages.[54]

XI.

As was the case with the New York Annenberg Challenge, we can use the notion of community to interpret what was occurring in Detroit during this reform initiative. Both the participation of community stakeholders on both the board of directors and council as well as the organization of the project in collaborative clusters were designed to promote a sense of common purpose among the participants. How successful the project was in activating the kind of civic capacity that community building requires is an open question.

At the level of the school clusters, it does appear that a number of mechanisms were put in place to promote civic capacity. The principal of one of the schools in Cluster # 57 commented about the Challenge that she "had a vision in my head about what I wanted. I always have from when I was a teacher... wanted the school to be accessible to the community... And I wanted to see adults, and I wanted to see kids offered more than just academics. I wanted their self esteem to be augmented. So I went ahead and felt like this was going to be an avenue for doing two things. It was gonna draw in the community and help us to relate to our community." She noted, in this vein, that the Challenge benefited not only children but adults. "It's brought," she pointed out, "adult education into the school—GED. I have seventy students enrolled in ESL. It brought two parent facilitators that go out into the community and work with parents and try to draw parent in and help us with, you know, problems with kids... I just think it's been supportive of refocusing the school and how it's managed, governed, and reaching out to our parents and the type of involvement we have with them. We have a coffee club every week and, you know, like we've brought in our parents." She went on to say that the program provided the school with the funds it needed "to reach out to the community."

The partnerships that the participating schools established with business and with voluntary organizations were especially helpful in addressing a variety of social problems that affected the neighborhood surrounding the schools. The teacher facilitator at Logan noted that the grant supported a full-time social worker from the Children's Aid Society stationed in the school. It also, she pointed out, funded an after school program that offered children numerous recreational and cultural activities that affluent families routinely provided but that the parents of children within the cluster could not afford.

The schools within the cluster were also able to hire parent facilitators. These individuals were typically community residents whose children attended one of the three participating schools. The facilities served as translators for parents who did not speak English. They provided parent workshops on all sorts of topics. At Logan, the parent facilitators offered among other things a family math night program, provided tutoring, and sponsored a coffee club for parents. Their role, according to one of Logan School facilitators, was "getting parent involvement in our school and trying to educate parents how to help [their] children at home."

As one of the parent facilitators saw it, their work had a marked impact on the program. She noted that at first parents of the children attending these schools seemed uninvolved. Their view, in her words, was "here's my child, educate him, bring him out, I want him educated." Over time, however, this facilitator claimed, that she and her colleagues were able to educate parents to look beyond their needs and understand the views of teachers and the schools. The result, she noted, was to increase the extent of parent involvement in the schools.

The employment of parent facilitators and the provision of space within schools for parents to meet were just two of the efforts that the Challenge undertook to better cement relationships with parents and with the community. The project supported an Ambassadors Program that enlisted parents, community and business representatives, students, local clergy, and school staff to promote the work of the Challenge in the course of their daily work and leisure activities. The project also sponsored a number of programs to promote reform. One such effort, "Stand Up for Our Children" was a campaign at the beginning of the Challenge to build community support for school reform. A second program, "Come to the Table," established a collaborative among students, parents, teachers, and community members to promote the role that the community could play in improving the city's schools. And a third project, "Achievement for All: Families and Community Working Together for High Standards," involved the Challenge and the school district in a joint campaign to inform the community about academic standards and the role that parents and the community could play in enhancing the academic success of the city's children.[55]

The key problem to obtaining civic capacity at the school level was the ending of the project. A parent facilitator at one of the Cluster # 57 schools made this point with respect to parent involvement. The culmination of the Challenge, she noted, will be "very unfortunate because we're gonna lose our parents. That's one thing for sure I can tell you. We are gonna lose our parents because with that money from Schools of the 21st Century, it has helped us educate our parents... buy them dictionaries, buy supplies so that they can help their kids at home, stuff like that. And it's gonna be a downfall. Yeah, I can see that."

Notwithstanding what successes school clusters, such as Cluster # 57, may have had in establishing a sense of common purpose among its stakeholders, there were problems involving the Challenge's intermediary organization, the Schools of the 21st Century Corporation, that undermined the emergence of civic capacity. One such problem was the fact that the Corporation was the only organization in the city promoting school reform. Unlike other Annenberg sites where there were several intermediaries promoting change, the 21st Century Corporation stood alone. It did not possess the kind of resources that were present in New York City where four organizations collaborated in their reform efforts.[56]

A second problem involved the relationship between the Corporation and the Detroit Board of Education. The Challenge appeared on the scene in the midst of the mayoral takeover campaign that we considered in chapter four. Neither the elected board that was trying to fend off the takeover campaign nor the new, mayorally appointed board and the interim CEO, both of whom were faced with developing a new administrative structure for the schools, were in a position to work collaboratively with the

Challenge. As a result, the efforts of the initiative often conflicted with the work of the board. At the same time that the Challenge was promoting curricular diversity by allowing schools to select the reform model that best suited their unique circumstances, the district was imposing a unified curriculum on all of the city's schools. In fact, in the last months of the Challenge, the board ceased to allow project schools to continue using reform models that departed from the city's uniform curriculum. It was often the case that principals of the project's Leadership Schools were placed in the position of having to choose between policies enacted by the Challenge and conflicting policies of the board of education. Curriculum was not the only issue in which the Challenge and the board disagreed. They also disagreed over the role that stakeholders outside the schools were to play in policy making and the authority that individual schools should have in managing their finances and other resources. As it turned out, much of the potential that an intermediary organization had for mobilizing reform was vitiated in Detroit by the conflicts between the board and the Challenge.[57]

At this point in time, it is unclear whether the efforts of the Challenge did in fact generate sufficient civic capacity to maintain this reform initiative. The project's final report points to the impetus that the Challenge has given to school improvement in Detroit. It notes in this regard that many of the external partnerships that the school clusters have established are being maintained without the funding provided by the project. And the report cites the establishment of what it calls its "legacy organization," the Detroit Parent Network, which is supposedly continuing the work undertaken by the Challenge to increase parental involvement in the city's schools.[58]

XII.

One of the dilemmas that partnerships pose comes with business participation in such ventures. It was the creation of a partnership between a for-profit company, Edison, and the New York City Board of Education that precipitated the charter school conversion conflict that we explored at the beginning of this chapter. Is it the case, then, that business participation in partnerships with schools dooms those enterprises to conflict? The work of the Minneapolis based intermediary organization, Youth Trust, suggests that this is not the case.

Organized in 1989 by the Minneapolis Community Business Employment Alliance, Youth Trust, which joined in 2002 with the Minneapolis

Public Schools Foundation to form Achieve!Mpls, represents a collaborative between the city's business community and its public schools to establish business-school partnerships.[59] The organization sponsors an array of programs that are designed to prepare the city's youth with marketable skills. The participating partners include a large number of the city's key business establishments, including American Express, General Mills, Pillsbury Company, and Honeywell. There are, however, participating partners from government and non-government organizations including Big Brothers and Big Sisters, the Hennepin County Attorney's Office, the YWCA, a local college, and several area churches.[60]

Youth Trust officials are quite frank about their commercial goals in establishing partnerships. According to the organization's staff member to whom I talked, they offer businesses "a good tax break, " and, "it is good public relations." At the same time, however, she noted that the participation of corporate employees in school partnerships represents "a true sense of volunteerism and shows that they care about the community that they're based in. It is all about making a difference in the life of a child [and] wanting to give back to the community."[61]

A Pillsbury executive who I interviewed noted the same mix of corporate and benevolent goals. "People," she commented, "go to the grocery store and they see the name Pillsbury, and they know that Pillsbury is active in the school system and it encourages consumers to purchase our products." She went on, however, to say that the partnership program supported those employees who live in the city and whose children attend the Minneapolis Public Schools. When Pillsbury employees volunteer in the schools, she pointed out, they learn more about the schools and come to realize that they are a good public investment. Volunteering also exposes their employees to a degree of ethnic diversity that they do not experience either in their workplaces or in the suburban neighborhoods in which many of them live.

During its ten years as an independent organization and currently as part of Achieve!Mpls, Youth Trust sponsors a variety of collaborative efforts between schools and business. In the city's elementary schools, Pillsbury offers a "Reading Buddies" program in which employees visit schools once a month and read with third grade students. In between these sessions, the Pillsbury employees exchange letters with their students. American Express offers a "Pen Pals" project in which company employees exchange letters with elementary schools students throughout the year. And employees from Dayton's Department store tutor students in a Minneapolis elementary school as part of its "Classroom Helpers" initiative.[62]

Among the middle school programs that Youth Trust supports are "e-Mentoring" and "Workplace Tutoring." In "e-Mentoring" employees from Norstan Communications and Cargill exchange weekly e-mails with middle school students, meet with the students in person several times during the semester, use e-mail to help them complete assignments, and provide them with information about the workplace and potential career options.[63] The "Workplace Tutoring Program," buses students once a week before school from two Minneapolis middle schools to offices of Honeywell and ReliaStar Financial Corporation where employees tutor them for one hour and then return them to school. The program is designed to provide students with help in mathematics and reading in preparation for state competency tests, to connect them with a caring adult, and to provide them with exposure to a work setting.[64] For high school students, Youth Trust provides its "High School Partners" program in which business volunteers are connected with high school teachers and students to develop projects to meet a host of unique classroom needs.[65]

XIII.

One of the often ignored contributions of school-business partnerships is as an instrument for linking corporations, schools, and neighborhoods in a network of common goals and purposes. The work of Youth Trust at Jordan Park Community School in North Minneapolis offers a good example. The school, according to its family partnership coordinator serves a "depressed, poor area." According to the 2000 census, the neighborhoods serving the school are about 50 percent African American with a little over half the families characterized as low income.[66] Of the almost 450 students who attended the school in 2004, 74 percent were African American and 89 percent qualified for free or reduced lunch. On the 2003–2004 Minnesota Comprehensive Assessment, the school's third and fifth graders did not attain adequate yearly progress (AYP) in reading or mathematics. Where approximately 70 percent of the state's third and fifth grade students reached proficiency in these areas, the percentage of Jordan third and fifth graders reporting proficiency ran from a low of 20 percent to a high of 34 percent.[67]

As Jordan's family partnership coordinator saw it, the challenge facing the school is that, in her words, "our kids are needing the support of all of the adults in the community." She went on to say that "it takes a village to raise a child and that is our purpose." The role of Youth Trust is to help to create partnerships that can fulfill this goal. One such partnership, Brain Trust, brings together corporate and community leaders to assist the school in resolving important issues. "We as educators," she noted, "think within our own box, and we were looking for people who thought out of the box... to come and be advisors or a think tank for us." The Brain Trust, which includes about fifteen individuals among whom are a vice president from U.S. Bank Corps, the chair of the Jordan Area Community Council, and the director of the General Mills Foundation meet once a month to help school officials look at issues in a broader way than they might consider them and bring resources to the school.

One such issue that she described involved deciding what to do with a monetary gift to the school from the General Mills Foundation. "The Foundation," she noted, "has taken the Jordan neighborhood under their wing [and] gave us a no strings attached gift of $10,000 to welcome us to the neighborhood." Working with the Brain Trust, it was decided to use the money to establish a scholarship program. Their recommendation was not for a scholarship for the most academically talented students but one for students who needed encouragement to get through high school to pursue further education. Piper Jafrey, who had a representative on the Brain Trust, matched the gift and with another $1,000 that the school raised, the scholarship fund reached $21,000.

In another partnership venture, the General Mills Foundation worked with the school to host a weekly Jordan Jam in which representatives from the school, business, police, and social service agencies meet with neighborhood residents to talk about issues involving community building. In addition, the school had entered into partnerships with Piper Jaffrey, Cargill, and the University of Minnesota to bring volunteers into the school to help students with their reading and mathematics skills.

From the vantage point of the family-partnership coordinator, these partnerships helped both the school and its surrounding community. "They bring in monetary resources and allow us to do things we could not normally do. They also bring in human resources, and it expands the community horizons by bringing in people to work with us and connecting us further with the community. It is all really for the benefit of the students because it gives us an opportunity to provide better teaching skills for our students. It benefits the community because they see that there are people who really care about the community and the children, and they really want to help to improve the community and the students."

Parents were also involved in these efforts. The coordinator described a $50,000 grant that the school received from Cargill, which was going to be used to improve students' mathematics achievement. One of the problems that the family-partnership coordinator noted was the difference between the way mathematics was currently taught and the mathematical understanding of parents. Mathematics today, she noted, is quite theoretical and does not stress computations skills to the degree that once was the case. She commented that this is "confusing to the parents because they believe that we should be teaching their children how to multiply, and they feel like we are not doing our job and are frustrated because they cannot help." Part of the Cargill grant, she stated, was going to be used to train parents so that could understand the curriculum and assist their children.

What was occurring at Jordan Park, as she saw it, were efforts like the Brain Trust to bring the school, business, and the community together. "Anytime you bring people in," she pointed out, "they become aware of the needs of the community and can bring financial help and support and hope and encouragement to the community... We want our students and families to see that they are a part of a broader community and this helps to stress that by bringing the groups together."

Despite the monetary and other resources that partnerships bring to schools like Jordan Park, it is often charged that such arrangements can enable the business sector to gain control of public education and use it for profit instead of public purposes. Jordan Park's family-partnership coordinator doubted that this would happen. "I think," she pointed out, "that one of the lovely things about involving business people is that it gives them a different point of view as to what is really happening in the school system, and when they come in, they discover the problems and needs are quite complex and that the people who are running the schools are doing the best that they can."

Finally, the experience of those who volunteer through Youth Trust also points to how school-business partnerships have the potential to bring diverse groups of people together around a common goal. A General Mills market researcher to whom I talked indicated that he became involved in a literacy program in the Minneapolis school in which his wife taught. His company, he noted, has a commitment to philanthropic work and encourages volunteerism among its employees. This latter encouragement was important to him, and he thought to other employees. He in fact, indicated, that he would not stay with the company if it did not provide him these kinds of opportunities. There was, he felt, a "good fit" between the needs of the schools and resources that General Mills had available. The program, as he saw it, helped "getting [General Mills] integrated with the community, being more in touch with the community." This was, in his opinion "community building." Similarly, the partnership helped connect the school and its neighborhood to the world of business. "It gives teachers, parents, and the kids," in his words, "more of a tie in and a perspective from the business community. They get to see our facility and relate to what we do. A lot of the students have no exposure to white collar America, and they need this."

A second market researcher who volunteered in the same literacy program had a similar view. "GM is," she commented, "a very community oriented organization, and it is very important to show that we are involved in the community." And this was precisely, again in her words, what the partnership program was doing in bringing "people from this company together to show the kids that different kinds of people care about one another."

A manager in a Minneapolis public relations firm who was involved in an e-Mentoring program with a Minneapolis middle school spoke of how the partnership served to establish the common ground between the corporate employees and students that supported the existence of a sense of community. "We were told by Youth Trust that one of the things that teachers have asked them to help with is to let these kids understand the things that they do in school everyday is for a greater purpose. They see you as people who are above them and have jobs that are elite, and it is our goal to say to them that you are working on the same things I worked on when I was your age." Another volunteer in this public relations firm made the same point. "We can," she stated, "bring something to the students that it would be difficult for them to get otherwise, whether it be another person that they can talk to about questions they might have about school [or] whether it be another person they look up to."

XIV.

In this chapter I examined the most recent effort of school reformers in New York City, Detroit, and Minneapolis to use partnerships as their vehicle for school improvement. Such initiatives, as we have seen, bring with them benefits and liabilities. They provide resources, human, material, and financial, that schools, particularly urban schools, could not otherwise afford. They have in this respect been a vehicle through which business has not only injected large amounts of both money and technical assistance to the schools, but have also promoted the task of improving urban schools to a wider public.[68]

In addition, partnerships can, as we have seen throughout this chapter but particularly in the school-business collaboratives supported by Minneapolis' Youth Trust, allowed for the kind of common purposes and goals that make community possible. In this vein, the educational historian Larry Cuban has noted:

> In mobilizing community elites to change schools business and school leaders using the media to spread their message, built social trust by bringing

> together parents, taxpayers, and neighbors to improve a critical social institution. In a process that scholars call building "civic capacity" and "social capital, business-inspired school reform in many—but not all—places glued together disparate segments of a city to act for the common good.[69]

Such efforts do not always work this way as the conflict between ACORN and Edison that we discussed at the beginning of this chapter illustrates. Yet, New York City and Detroit's Annenberg Challenge and Minneapolis' Youth Trust program do point to how partnerships can be vehicles for creating a sense of collective belonging among a diverse urban population.

Partnerships do have their downside. As we saw in this chapter, private contributions to schools, either in cash or in kind, can have a limited longevity and the withdrawal of support can stop reform dead in its tracks. Private partners, it is often the case, have goals and agendas of their own that are not necessarily shared by educators and can compel schools in return for support to take on programs and policies to which they do not necessarily subscribe. We also saw in this chapter that the actual record of partnerships in enhancing academic achievement has been at best mixed. Critics of this reform argue that this less than successful record points to their mistaken emphasis on addressing the problems of urban communities by infusing resources into schools to improve their pedagogical practices and curricular programs. Such initiatives focus their attention on unleashing the individual efforts of urban youth from the blocked opportunities of inadequate schooling but ignore the more fundamental structural problems of American society rooted in poverty, racism, inadequate housing and health care, and employment discrimination.[70]

Critics also note that partnerships involving business can raise the specter of privatization, which in turn leads to the kind of conflict that surrounded the effort of Edison to pull off a charter school conversation in New York City. The involvement of business in such partnerships has been particularly contentious and brings with it an array of charges to the effect that business is a corrupting force in public education that subordinates schools to the logic of neoliberalism and the market. Schools, according to this viewpoint, become venues for marketing commercial products, exploiting teachers and students, and transforming an institution designed to promote the public good of the many to the profit motives of the few. Such criticism goes further to see partnerships and other forms of business involvement in the public schools as an indication of the power of corporations in a globalized world to undermine the institutions of democracy.[71]

Partnerships as reform vehicles clearly do not exist in a void. Their recent popularity can be linked to the impact that forces of globalization are having on the ability of the state and its agencies to support such social provisions as education. What is not as clear, however, is whether their impact is as negative as the critics of these ventures claim. I will explore this issue in the next chapter by moving across the Atlantic to consider the attempt of Britain's New Labour Government during Prime Minister Tony Blair's first term in office to employ partnerships to remake the country's state system of schooling.

Chapter 6

Educational Partnerships and Community: Education Action Zones and "Third Way" Educational Reform in Britain

Urban reform is clearly not an American invention. In his masterful study of Progressive era reform, Daniel Rodgers has located the roots of many of the initiatives for social betterment that were a part of this early twentieth century movement in Western Europe, particularly in Germany and Britain. In his volume he describes something of a movement across the Atlantic in which Americans took their cues for constructing the reform practices of the Progressive Era from European policies and programs in such areas as social insurance, housing, and city planning.[1] Over the course of the twentieth century and now into the twenty-first this movement of reform policies and practices across the Atlantic has ebbed and flowed as well as shifted directions in response to various social, economic, and political transformations of the moment.

The reform policies that we have explored thus far in this volume have not required us to look outside the United States. This is less so the case with the discussion of partnerships that we began in the preceding chapter. In this instance, we will find that looking at how policy makers in Great Britain during the last three decades have made use of partnerships in educational improvement provides us a fuller and more complete picture of this initiative. It was Britain's post–World War II economic dislocations that provided the impetus for the appearance on the scene of partnerships as reform vehicles. Neither the Keynesian policies of Labour governments during this period nor the less interventionist actions of the Conservative Party seemed able to deal with the country's pattern of slow economic growth, high unemployment, and high inflation. The demand that the International Monetary Fund imposed on Britain during the 1970s to control public expenditures in return for loans to deal with the country's economic problems led both political parties to seek alternatives to the direct state financing and operation of public services.[2] Margaret Thatcher's inclination, when the Conservatives came to power in 1979, was to look to America for possible solutions to the country's public service dilemma. One U.S. reform for social provision that caught her interest were the compacts that had been established between business and school systems to link the provision of jobs and private sector training opportunities for students to efforts on the part of schools to raise achievement levels. It was the visits of Conservative ministers and policy makers, first to New York in 1984 and then to Boston in 1986, that led to the formation of an array of similar business-school compacts or partnerships in London, East London, and a number of other cities throughout Britain.[3]

Despite the fact that the Conservatives were the first to embrace partnerships in the United Kingdom, it was in the Labour government of Prime Minister Tony Blair that was elected in 1997 that featured partnerships as the center piece of its educational reform agenda. There are several reasons why Blair found this notion of partnerships so attractive. The idea of a partnership seemed to fit nicely with his stated communitarian leanings that stemmed back to his student days at Oxford and his exposure to the philosophy of John Macmurray, particularly his understanding of the notion of community.[4] Further, Blair's commitment to "third way" thinking, which with its penchant for reducing state direction of economic and social policies seemed to lead almost inevitably to an embrace of partnerships.

"Third way" thinking had emerged at the beginning of the 1990s as part of the effort of a host of politicians and policy makers in the United States, United Kingdom, and Western Europe to respond to an array of economic, social, and technological changes that have occurred during the last three decades of the twentieth century that, according to the sociologist Anthony Giddens, "cut across the borders of nations."[5] Labeled as globalization, the most important of these transformations for theorists of this movement has been the appearance of immediate world-wide communication and transportation systems. Other changes, all of which require global communication and transportation networks include the following: the expansion of trade and investment across national borders, the growth of international financial markets, the free movement of labor across national borders, the immediate impact of distant events, the global effect of local economic and political decisions, and the redistribution of power among nation states, local entities within nations, and regions that span national boundaries.[6]

As it turns out, globalization is a contentious concept. There is disagreement about whether such a process has ever occurred or is occurring as well as the specific changes it supposedly encompasses.[7] Yet, most proponents of "third way" thinking accept its reality and argue that such changes have rendered the modern world less stable, less secure, and more subject to immediate change.[8] They go on to advocate three policy initiatives to address this more fluid world. First, they are committed to a reconstruction of the state that promotes the devolution of authority and the decentralization of decision making. Second, its proponents wish to enhance the role of associations and other vehicles of social capital within communities in civic betterment and see joint ventures among civil society, business, and government as the best means for achieving this goal. They often speak of this goal as restoring a sense of community and a commitment to the common good.

Third, supporters of the "third way" see the relationship of the individual to the state as a collaborative built on mutuality and reciprocity. They speak of this connection in terms of "no rights without responsibilities." In other words, such benefits as unemployment insurance and welfare that individuals receive as part of a social safety net are not simple entitlements. Such benefits are temporary measures that entail an obligation on the part of recipients to actively seek and accept paid work. Finally, advocates of this viewpoint seek to make society more equal or more socially inclusive. By this they mean enhancing the access of individuals to both the rights and responsibilities of political citizenship and the opportunity to participate in the life of society. Their strategy for achieving this end is to promote individual access to what they see as the vehicles of inclusion, namely education and jobs. The more equal society that they favor, then, is one that offers opportunity and the success and failure that result, not equal outcomes. It is for that reason that "third way" politics recognizes the need to provide for those who are less successful with the proviso that they balance rights with responsibilities.[9]

Partnership is the term of choice among a number of "third way" proponents, including Blair, for describing the relationship that they envisioned between the public and private sector under conditions of globalization.[10] In a 2002 address at the annual Labour Party conference, Blair spoke of partnerships as "the antidote to unilateralism," as "citizenship for the 21st century," and as the vehicles for the delivery of the services that society provides to its members.[11] Seen as an antidote for historic battles between left and right and public and private, a partnership is a way of bringing together government, business, the voluntary sector, and citizens around issues of public policy. It is in other words a mechanism for cooperation and consensus. Partnerships, then, constitute an alternative to prevailing notions of regulation, to both the controlling role that liberal and other left-leaning thinkers have bestowed on the state or government and the similar role that conservatives and neo-liberals have given to the market. Partnerships offered New Labour a strategy that would enable the party to maintain its longstanding tradition of supporting public services while at the same time allowing it to maintain its new stance of controlling public expenditures and reducing state intervention.[12] In this chapter I will explore their most extensive use of this approach to school improvement as exemplified in their high profile but short-lived Education Action Zone initiative.

II.

First established in 1998 by Britain's New Labour government, Education Action Zones (EAZs) represented a key initiative in Prime Minister Blair's effort to build a politics reflecting one of his most-cherished values, community.[13] These zones brought together clusters of usually fifteen to twenty-five schools located in areas of social and economic distress with the intent of raising academic standards in low achieving rural and urban schools thereby enhancing the social inclusion of the population. They were collaboratives involving schools, local education agencies, parents, community groups, and the private sector and were funded by a combination of direct government support and private sector, particularly business, contributions. The driving force behind these zones was the establishment of partnerships among individual schools, parents, business, community organizations, and the voluntary sector. Business was to play an especially important role in the management of these zones and in their financial support. The government sought applicants from a number of sources including individual schools, businesses, parents, and community organizations. The Local Education Authorities (LEA) that were the local administrative unit in managing Britain's system of state schooling did not have to be involved but could be one of the partners in the management of a zone. Ultimately, there were to be seventy-three EAZs throughout Britain.[14]

As it has turned, out the life of this initiative was short. In November of 2001, three years after the then school standards minister, Stephen Byers labeled the zones as "the test bed for the education system of the 21st century [and] a fundamental challenge to the status quo," the New Labour government announced the program was being disbanded and that none of the existing zones would receive funding beyond the original five year commitment.[15] By the end of 2004, then, EAZs in their original form ceased to exist. They did not, however, exactly disappear. Within a year of introducing its original EAZ initiative, the government put forward another proposal for a number of smaller EAZs as part of its Excellence in Cities initiative for addressing low achievement in inner-city schools. These zones like their larger counterparts were expected to develop partnerships with external stakeholder and to seek private sector support. What was different about them, however, was that LEAs provided the government share of their funding and also administered them.[16]

At the heart of New Labour's political agenda was the goal of restoring a sense of community to modern British society. Since the early 1990s, Blair and his supporters have invoked a number of terms and phrases to capture the central goal of their movement within the Labour Party and to differentiate themselves from the then reigning Conservative Party. These notions have included, among others stakeholder society, one nation, partnership, and community. Illusive terms at best, they point in somewhat different ways to what Blair saw as Britain's long standing but currently absent values of "social unity, common purpose, fairness, and mutual responsibility."[17]

When New Labourites used community or these other notions they were referring to a politics that sought to reconcile the antagonisms that have existed in Britain since the end of World War II between any of a number thought to be extreme positions that have been variously labeled as public and private, left and right, government and industry, state and markets.[18] Education Action Zones represented an initiative then to link the government's education goals of raising standards in low performing schools in disadvantaged settings with its broader political goal of building community.

III.

Coming to power in 1997 after almost twenty years of Conservative rule, the Labour Party saw educational reform as its first priority. Key to this agenda, according to its White Paper on education, were initiatives to ensure that schools held students accountable to high academic standards. Outcomes, it was believed, were more important than how any particular school was organized or structured. There would be, the report goes on to say, no tolerance for failing or low performing schools. Such schools would have to improve or be closed. The emphasis on high standards, according to the White Paper, must apply to all students. Schools had to do more to ensure the fair and equitable treatment of children who had been excluded from the benefits of education, particularly ethnic minorities. In the report, the government reaffirmed its commitment to equal educational opportunity but noted that it would modernize the principle of comprehensive education by establishing a number of specialist schools in technology, languages, sports, and the arts to meet the diverse talents and interests of children. It was, they argued, a means of making schooling more inclusive while not ignoring the very real differences among children.[19]

If these changes were to occur, New Labourites argued that it was necessary to change the existing governing practices of the schools, particularly the role of Local Education Authorities. LEAs traditionally had broad authority in providing local control and regulation of schools in England and Wales. They had control over curriculum, budgets, and the hiring of staff. The Conservative government that came to power in 1979 began to limit the authority of LEAs on the grounds that they had not been particularly successful in raising academic standards in the schools, a task that they transferred to the national government through the newly created Office for Standards in Education (OFSTED). To a large degree, the Blair government continued to support a lesser role for the LEAs for the same reason as did the Conservatives but has called for a recalibrated relationship between these agencies and the schools. LEAs would no longer directly govern schools but would serve as a mediating institution to enlist schools, business, voluntary agencies, and LEAs themselves in the work of increasing academic standards.[20]

Partnerships represented one of an array of initiatives that New Labour introduced to chart its own course to reform, one that Blair and his colleagues argue differentiates them from earlier Labour governments. At the heart of that difference was their redefinition of the concept of equality. When past Labour governments invoked this notion, they were referring to the task of equalizing the conditions of life among the citizenry. A combination of policies including increased public ownership, a redistributive taxation scheme, an extensive system of public welfare, and a commitment to full employment would have the effect over time, Labourites claimed, of diminishing the status differences and outright unfairness that they associated with Britain's class structure.[21]

Equality for New Labour meant something different, namely equality of opportunity. New Labourites explicitly rejected redistributive policies that sought to reduce the economic differences that separate the rich from the poor in favor of efforts, which they claimed would enhance everyone's chances to succeed. Invoking notions of community and the "stakeholder society," Blair envisioned a Britain in which all its citizens have a share in the fate of their nation and are bound together in a relationship of "social unity, common purpose, fairness, and mutual responsibility."[22] It was an approach to governing that its promoters saw as avoiding both the inflexibility of old Labour's statism without falling prey to the vicissitudes which they believed have accompanied the Conservatives faith in markets.[23]

The route to equality for New Labour was through social inclusion, which in their lexicon referred to access to education and work. This was not so much a right for New Laborites as it was an obligation. In return for their ability to participate in society, individuals had the responsibility to look for and obtain paid employment and to acquire the skills and knowledge that they need to be prepared for work. It was not, however, the earnings that made work inclusive. Rather it was the sense of self-esteem and self-worth that being employed held out for individuals.[24]

The importance in this shift in language becomes clear when we look at how it played itself out in policy, specifically educational policy. The problem of low academic standards and low achievement was not for New Labour a structural problem resulting from economic inequality, class bias, or racism. Rather, it was a problem of deficient families, especially parents whose own history of unemployment, poverty, and lack of education has left them without the skills, self-esteem, and ambition to help their children succeed in school. New Labour's remedy for this problem was not the traditional Labour policy of redistributing resources from successful families to less successful ones. What they favored was a combination of initiatives that offered parents the education that they needed to obtain employment, which in turn, they claimed, would provide them with the desire and sense of self-worth to engage themselves in their children's lives and schooling coupled with policies that compelled such parental involvement.[25] The way forward in education, for New Labour, was in effect a partnership between parents and the state to provide them with the knowledge, dispositions, and skills that would make them employable and this in turn would place them in a position to advance the educational success of their children. They also favored partnerships between families and schools in which parents assumed responsibility for ensuring the educational success of their children.

IV.

It was through the establishment of Education Action Zones that New Labour sought to use partnerships as their mechanism of reform. The EAZs came with several incentives to make them attractive to schools and other potential applicants. The first twenty-five zones received £750,000 ($1,200,000 [1998 exchange rate]) of government funding annually for an initial three year period with the prospect of extending the life of the zone and its support up to five years with satisfactory performance. Each of these zones was expected to supplement this government funding by raising £250,000 ($400,000) in cash or in kind annually from business or other private contributions. The forty-eight zones that were established subsequently received annual grants of £500,000 ($800,000) and matched funding up to £250,000 for each pound sterling of private sector funding that they obtained.[26] Beyond additional funding, Education Action Zones were to be allowed priority in introducing any of a number of other government initiatives. These programs included efforts to enhance the quality of teaching and learning, to provide support for families and students, to partner with external organizations, and to promote policies of social inclusion.

Secondary schools within an EAZ were given preference when applying to the government to become specialist schools, which provided them additional public and private funding to support a curriculum that included foreign languages, technology, or sports. Zone schools could also establish links with Beacon schools, existing schools that were allocated additional funding to enable them to use their know how to aid other schools. They could introduce curricular innovations that would depart from the National Curriculum. And they could also introduce more flexible conditions of employment and alternative salary schemes that varied from the National Teachers' Pay and Conditions Document.

The administrative unit for a zone was its Action Forum, which served as a venue for bringing together the participating partners. Although those submitting applications for a zone developed a specific administrative structure for their forum, they typically comprised representatives from participating schools, parents, businesses, community and voluntary organizations, and the local educational authority. A forum, in turn, appointed a project director who assumed responsibility for the day-to-day management of the zone. Once established, a forum could assume any of a number of roles. It could leave the running of the schools to their existing governing bodies and focus its attention on raising standards and other overall zone targets. A forum could, however, serve as an agent for one or more participating schools in carrying out specific zone responsibilities. Or a forum could, if participating schools were willing to relinquish authority to it, assume responsibility for most of the functions of a zone and become the EAZs single governing body.[27]

There was an array of critics of this initiative outside of the New Labour government, especially university academics. Much of the criticism was there from the start and was not directed exclusively at Education Action Zones but more broadly at New Labour's "third way" oriented educational policies. For these critics, the government's educational program was to a large extent a warmed over version of the neoliberal, market oriented approach of the previous Conservative government with its penchant for privatization and competition over comprehensive state schooling and equality.[28]

They also challenged the record of EAZs in enhancing student academic achievement. The actual results, they claimed, were uneven and inconsistent with zone schools both surpassing and falling below national levels of attainment and those of non-zone schools within their LEAs.[29] They also noted that increases that were occurring in academic achievement within EAZs were also occurring in non-zone schools and that consequently the gap between different segments of society was not declining. EAZs were not, in other words, the vehicles for raising standards that their proponents claimed.[30]

What especially bothered these critics about EAZs was the threat to state schooling they saw from the reliance on private partnerships for the funding and management of schools. Such collaboration, they feared, could ultimately undercut public control of the nation's education system. In that vein, they saw the zone's ability to alter the salary and working conditions of teachers as an attack not only on teachers themselves but more broadly on working people throughout Britain. And similarly, they thought that the authority given to EAZs to depart from the National Curriculum and to establish specialist schools could undermine comprehensive education in favor of a narrow, vocationalized course of study that would channel children of the poor to decidedly unequal occupational and social roles.[31]

At the same time, there were critics of the initiative who questioned the actual efficacy of partnerships with business. They noted in this vein that business contributions never reached their expected levels, that the money that was contributed often came from the voluntary and public sectors, and that much of the contributions were in-kind rather than in cash. There were, they went on to report, instances in which private sector cash and in-kind donations to LEAs and not specifically to individual zones were counted anyway as part of the EAZs private match. It was also the case that business involvement in the management of zones was less than expected. These critics noted that there were not that many business partners who had any interest in operating EAZs and that the participation of business in the Action Forum was often minimal. As it turned out, LEAs were central players in the establishment and management of the EAZs notwithstanding the government's desire to reduce their role in favor of the private sector.[32]

Critics were also worried about the partnerships between schools, parents, and communities that were promoted in the EAZ initiative. These were not, they claimed, equal partnerships where parents and communities were to play a role in the management of zones. They were, again to their way of thinking, clearly unequal relationships that sought to use collaboration to enhance the control of school authorities over parents and communities. What so concerned these opponents of EAZs was the government's assumption that low achievement was not a structural problem but rather the result of deficient families and communities and that the goal of this program was to repair these deficits. In this vein, they were particularly critical of what they viewed as the coercive and authoritarian impulse within this initiative to virtually compel parents to assume principal responsibility for the education and ultimately for the employability of their children. The commitment of the zones to social inclusion did not, they argued, mean that the government sought to use this initiative to achieve equality. Rather, they went on to say, it was an approach, rooted in human capital theory, to make the poor and disadvantaged more employable thereby enhancing the country's competitive advantage in a global economy.[33]

And finally, there were those commentators who questioned whether it was even possible to distinguish the true impact of the EAZ initiative from New Labour's assertions about its supposed successes. As they saw it, the claims about this program were often caught up in a process of "spin" or "impression management" that obscured its actual impact behind a rhetoric emphasizing the government's seeming problem solving acumen or insight.[34]

The government challenged these criticisms. They claimed that the EAZ initiative has been a success in raising academic standards, reducing gaps in levels of achievement among British youth, securing business involvement in the financing of zones, and promoting new governance schemes, particularly partnerships with business, parents, and the community.[35] One member of the EAZ team in the government unit that administered the program in Blair's second term, the Department for Education and Skills (DfES), denied that the lack of private funding had anything to do with the demise of the program. From the beginning, he went on to say, New Labour officials did not see Education Action Zones as permanent entities. They were, he noted, temporary initiatives that brought with them a number of important successes that would constitute the basis for further educational reform. As the government saw it, he pointed out, the best strategy for the future was to integrate this program into their overall school reform strategy, which was being done through the Excellence in Cities initiative.[36] Another member of the EAZ team noted that financial support was not the most important contribution that business brought to this reform effort. "The cash is useful," but she went on to say that "these businesses have specialties in more general skills such as management of staff, that sort of thing. We wanted to draw on all of that, and the cash is very nice, but it isn't really the biggest thing."

V.

To see how this effort to build a sense of community played itself out in practice, we will explore the experiences one of England's most disadvantaged communities with the EAZ initiative, an area I am calling the Borough of North Upton.[37] Established in 2000 and comprising eleven of the community's fifty-eight primary schools and five of its nine secondary schools, the North Upton EAZ was located in the kind of disadvantaged and distressed area that the government had in mind as a site for this initiative.[38] A borough within London, this community has over the last two decades suffered from a combination of political party discord, financial mismanagement, and administrative incompetence. Recent problems included a large budget deficit, the inability to collect local taxes, the loss of key civil servants through attrition and cut backs, the deterioration of the public housing infrastructure, and the collapse of such essential services as trash collection.[39] North Upton, according to the head of one of the primary schools in the EAZ, has been a "dysfunctional borough for all the years that I've been in it and words fail me to know...how bad it has been."

Recent immigration has transformed this once largely white working class area of London into a racially diverse community with large numbers of recent arrivals from Africa, the Caribbean region, and Turkey. Poverty and unemployment were key problems facing the residents at the moment. North Upton's average gross household income in 1993 was £11,900 compared to the average gross household for the city as a whole of £19,700. The unemployment rate for the community in 1999 was 14.7 percent, which was almost three times higher than that of entire City of London. Other community problems included inadequate housing, high levels of infant morality, and high incidence of violent crime.[40]

Not surprisingly, these larger community difficulties have affected the schools. The student population within the schools comprising the EAZ was 75 percent ethnic minority. The percentage of these students eligible for free school meals and whose native language was other than English exceeded national averages. In fact, children who attended schools within this Action Zone spoke over eighty languages. The schools comprising the EAZ exhibited patterns of persistent low achievement. The average performance of primary students was below the national average in reading, mathematics, and English, and the percentage of secondary school students who scored adequately on mandated competency tests was below the national average. These deficits were particularly severe among students of Turkish and Caribbean descent.[41]

The LEA that managed the schools since the early 1980s experienced many difficulties similar to those that have affected the larger community. One primary school head teacher defined the problems of the local authority as "incompetence, corruption, nepotism." It was, he went on to say, "the most troubled education authority in the country bar none." The LEA, according to this administrator, was particularly inept. Many of those who were hired when the authority took over control of the schools were put there because of their political influence. Positions were given to individuals who had neither training nor experience in education, including former clerks and recent immigrants without work permits. The LEA did not fare well in the educational reform environment of the 1980s and 1990s. OFSTED inspections in 1997 and 1998 pointed to a number of key failings on the part of the authority including its inadequate budgeting processes, its seeming inability to support failing schools, a poorly designed education development plan, and its lack of success in introducing information technology into the schools.[42]

David Blunkett, the Education and Employment Secretary during New Labour's first term, recommended that some of the LEAs services be contracted out to private businesses, which it was thought could operate them more efficiently. This first effort at privatization brought some slight improvements in student academic performance. Yet, continuing management problems led to a third OFSTED inspection in 2000, which concluded that the LEA continued to function at an ineffective level and did not appear capable of achieving much in the way of improvement. Two years later a government committee recommended that private not-for-profit company take over the management of the borough's schools from the LEA.[43]

Despite North Upton's array of educational and social problems, the establishment of an EAZ was not exactly a welcomed event. The only teacher's strike in the country in opposition to the initiative occurred in one of the schools in the borough. The local teachers association was a key opponent. As one of the association's officers noted, "the problem with Education Action Zones, first of all, [is] they're unfair. They provide additional resources, which should be going to all schools, to only a few. Secondly, such resources as they do provide are heavily into the administrative costs of running them. And this particular one, for instance, has a director, has a deputy director, has administrative support. And it seemed to us neither a fair, nor an efficient way of providing real funds that all schools need. And finally... their aim was to attract in Tony Blair's third way fantasizing capital from the private sector, which they seem to [have] failed to do." More troubling to his organization than the EAZs seeming lack of efficiency was their fear that the business partnerships promoted in this initiative would lead to the privatization of public education. As he put it, "we don't want the curriculum brought to us by McDonald's, thank you very much indeed."

The fear of business control was a major source of teacher union opposition. Union members did not trust the intentions of private business. Equating the EAZ business partnerships with the privatization initiatives of such U.S. firms as Edison, they saw business support of and participation in the management of the zones as part of a scheme for deriving profits out of operating public schools. The union official interviewed in North Upton saw business involvement in the EAZs as a minor phase of a larger corporate effort. In his words, "what they're 'buggering' around with here is for marginal profit." Their real mission was to position themselves "for very big global stakes" as the providers of educational services and products for the developing world. The EAZs offered them a place to "brand their product."

Commercial exploitation was only part of the problem that teacher organizations saw in the kind of privatization brought by the EAZ program. The governance structure of the zones, particularly the transfer of authority from LEAs to Action Forums threatened the kind of democratic management that they saw as the hallmark of a system of public education. The authority these zones had to depart from the National Curriculum could at the behest of a business oriented administration lead to the introduction of a narrow vocationalized curriculum that channeled students to decidedly unequal occupational and social roles. This control over the curriculum coupled with the power of the zones to establish specialist secondary schools could allow EAZs to undermine the country's ostensible commitment to comprehensive education. Finally, the unions saw the ability of the zones to alter the conditions of teachers' work and pay as one phase of a larger attack on the country's working classes.[44]

The unions were not the only opponents of the EAZ in North Upton. A primary head noted that she experienced some opposition from her governing board. "The governing board was skeptical here. They didn't believe that the money would actually be forthcoming. They felt that the sorts of activities that we might involve ourselves in would detract from our core task of education and instead of being clear about getting all children to certain educational levels, we might be involved in interesting and exciting or different or weird kinds of subject which would detract. They weren't keen on that. They felt that it would take up a lot of my time and energy and so I wouldn't be able to devote myself to what I should be doing. They were unhappy that somehow their powers might be ceded to the forum." Parents at this same school also had initial doubts about the EAZ. They were afraid that joining the zone would force them to share their resources with poorer and less well performing schools or that their children would be bussed out of their school to schools elsewhere within the zone.

In North Upton, as it turned out, the opposition to the EAZ was unsuccessful. Head teachers, focusing almost exclusively on the money that the program would bring to their schools, pushed their staffs and governing bodies to approve their participation. One primary head teacher commented that "I had lots of meetings with staff, lots of meetings with governors, lots of meetings with parents, and in the end I took quite a hard line, which is to say in my professional opinion and in my judgment, this is a good thing, and I fully believe that we should be doing it. It will be of benefit for the children and the community. And I almost dared them to go against me and a lot of parents, I think, were very comfortable, because they trusted me and they said okay if [I think] this is okay, we'll go for it." In the end the governing body was divided. Half of them supported the head in her decision to join the zone and, one or two resigned.

Another primary school head reported that he was put off by what he viewed as the "bizarre" charges of the "militant union people" who were leading the opponents. "The more I heard the opposition, the more I thought it was a good idea." There were, he noted, two factors that led him to try to convince his staff to support joining the EAZ. One was the money that it would bring into his school. The other was his confidence in the individuals he met would lead the zone. "The thing I liked about them was that they weren't kind of 'waffley'." They were very much this is what we want to do. Do you want to come on board? This is what you get for your school. You have this much autonomy to do you know what you want with it, and this is what we'd like from you."

VI.

As it has turned out, the impact of Education Action Zones on academic standards was mixed. The zone's director noted that both academic achievement and attendance in the secondary schools increased, but at the same time achievement in the primary schools "remained static." A primary head teacher supported this assessment. As he saw it, however, the problem was with one or two schools "that consistently failed and dragged the aggregate down." He went on to say that if we leave out the schools that were failing, then the EAZ had "a very big impact on standards." It is important to note, he continued, that at the same time that standards were increasing in North Upton, they were also rising throughout the nation. The head teacher in one of the zone's secondary schools also noted improvement in student achievement in English and science along with a small, but not to her way of thinking significant, drop in mathematics.

Where the EAZ did make a significant impact in North Upton, however, was on the curriculum and programs that participating schools were able to offer. As a primary head teacher saw it, academic standards were not the whole story. "It's the fantastic opportunities that it gave children... to learn French, to have emotional and behavioral counselors, to have fantastic ICT (information and communication technology) provisions—that's what the EAZ was all about." At one primary school, EAZ funds supported bringing professional musicians to the school to promote student involvement with music and to offer the children the opportunity to learn French. At another primary school, funds were used to employ a part-time sports coach, to install interactive white boards in all the classrooms, to construct two ICT suites, to host visiting poets, and to employ counselors.

According to this head teacher, the funding that came with the EAZ program supported what he called the "peripherals." "It provided some of the nice ICT stuff, some of the nice mentoring stuff, the counseling stuff; all that nice stuff on the side. The impact on literacy and numeracy, I personally have to say I can't see what it is." The head teacher at one of the community's secondary schools also saw the initiatives that she had undertaken with EAZ funds as separate from what she was doing in the area of reading and mathematics, which she pointed out, was supported by additional funding apart from the EAZ initiative. The programs that she undertook, including the establishment of parental partnerships, support for staff development, improving ICT facilities, and the introduction of vocational courses were, she noted, largely "revision and enhancement activities outside the normal school day." Yet, at the same time she saw them as fitting within her overall strategy of "raising attainment as well as raising opportunities for students."

As the head teacher at one primary school pointed out, "there is your core curriculum, which we have to deliver—the National Curriculum—and other bits you can attach to that but just the possibility to widen it for the children, and I think it has been our biggest achievement." She noted that her school was one of the few schools in the community to offer French and that "we've been able to take off with ICT in a way that I'm not sure we would've been able to if we didn't have, if we weren't part of the EAZ." A parent and member of the Action Forum at another of the borough's primary schools noted how EAZ funds had allowed the head teacher to work collaboratively with a local secondary school to provide her daughter with work in science, mathematics, and English that she had not been getting at the school. She went on to say that through this initiative a mentor from a local business worked with her daughter every Wednesday on mathematics and as a consequence her work in this area improved. She also noted that the program supported a partnership with a local football club that has allowed students, such as her son, to participate in a school sponsored soccer program. The EAZ provided, she stated, that "extra something that's been really good for them, and also it's helped the secondary transfer because my daughter said she wants to go to the local comprehensive because she been there so often because of the EAZ."

The promoters of the Education Action Zone initiative, we have already noted, saw the enhancement of academic achievement as a strategy for promoting social inclusion. Increasing educational standards would, in other words, provide the populations of economically distressed areas with the work related knowledge and skills they required for participation in the national economy.[45] Communities that sought to establish EAZs were required in their applications to show how in good third way fashion their proposed zones would address problems of low academic achievement and economic distress by fixing certain parental and family deficits.[46]

Repairing deficient families, then, was a common theme in EAZ applications throughout the country. The South East Sheffield application noted that the city's high levels of unemployment and poverty were the result of intergenerational factors that contributed to the community's low level of "self-esteem, educational aspirations, and motivation." The principal objective of their proposed EAZ was "to break the cycle of depressed aspiration and self esteem and to lift the locality's attitudes towards education and individuals' belief in their capacity to succeed and to move on." The route to achieving this goal, the applications stated, was to make parents more effective in supporting their own learning and as well as that of their children.[47] Similarly, the application from Hastings and St. Leonards posited a direct link between family deficits and economic and social distress. "Many families," the applicants argued, "have poor health and low self-esteem due to socio-economic factors. Education is perceived to be a low priority; therefore aspirations attendance, attainment and...skills acquired for employability are also low."[48]

In the same vein, the North Upton application noted that the zone would enhance the academic achievement and ambition of the borough's children by "targeting parental and community involvement in education." The EAZs proponents went on to say that "economic regeneration should be driven by sharply rising educational standards. Confidence, self-esteem, high aspirations and good educational standards will open up employment opportunities for... young people in local business and in the City and the rest of London."[49] As the zone's director noted, "quite a few women do... cleaning, waitressing, and so forth, for the big city firms, but the great raft of administrative, secretarial, executive jobs in the financial sector are not done by [North Upton] residents." He went on to say that "the city does not employ in the mainstream jobs of the financial sector a larger percentage of [North Upton] students, and [North Upton] students, in a sense, feel that isn't for them, and on the other hand, the city is saying you're not for us. So the Action Zone is a deliberate attempt to further break down those barriers."

A banker who served on the North Upton Action Forum made the same point when he talked about the "virtuous circle" that the Education Action Zone could create between the community's inhabitants and the city's commercial and financial sector. "If we can get business involved in education, young people who may live in conditions of hopelessness domestically, will see a glimmer of hope and as a consequence will raise their ambitions and hopefully their achievements... As academic achievement rises, employability rises. As employability rises, wealth is brought into the micro-economy; that will in turn allow local companies to have greater demand for their services and hopefully spawn more local business opportunities and so on." The zone, as its proponents saw it, would not only increase academic standards. It would instill the community's youth with other attributes, particularly, "confidence, self-esteem, and high aspirations," that taken together would have the affect of opening up opportunities for these young people in business in both North Upton and London. The plan, according to its proponents, was predicated on the belief that the best way to enhance student achievement was to ensure that parents were actively engaged in supporting both their children's education and their own.[50]

Despite the intention of the promoters of North Upton's EAZ, it is not certain that the emphasis on social inclusion found its way into the schools. The head teacher at one primary school indicated that she had plans to keep the school open from 8 am to 6 pm and to provide breakfast and a range of before and after school activities for children from families in distress. "I want us," she noted, "to be picking up children much earlier, so that we're intervening in their education and giving them the support that they need from a much earlier age and building on that through the school." She went on to say, however, that at the present moment she did not think that the EAZ was having any impact on the borough's economy. She mentioned some "big plans" two years down the road to make their computer facilities and library accessible to parents, which "might give them skills to become employed." These were, she felt, small efforts that were just beginning.

VII.

Educational Action Zones were not, for New Labour, just another educational innovation directed at low achievement and social disaffection. They were to be innovative. Most important in this respect was that they were to be partnerships that linked the schools to parents, business, the community, and voluntary organizations in the work of school reform. They were designed to promote new schemes of finance and governance involving substantial contributions from the business sector to support their operation and alternative administrative structures. And to allow these managerial reforms to spur forward change, the zones were given authority to operate outside the framework of both the National Curriculum and national regulations governing the salaries and conditions of work of teachers.

North Upton's record of implementing these innovations, however, has been mixed. The zone was clearly successful in establishing partnerships with an array of business, community, and voluntary agencies. The cash grants that came from these partnerships were not, according to the EAZ director, "massive," but the combination of smaller grants from numerous business and voluntary sector agencies meant, according to a primary school head that "there was a lot of private money that came in, a lot of initiatives," and he went on to say that "I am happy to take it." A secondary school head noted that it was the support of the EAZ and their links to business that enabled the school to secure the funding necessary to take on specialist status and become a sports college. She also noted that the school was able to refurbish its ICT suite with a £30,000 ($48,000) donation from a London based company. And the EAZ itself provided funding to support a corrective reading program for her low achieving Black students and a language instructor to provide help for Turkish students. A primary head teacher reported that over the course of a week she was able to garner £1,400 ($2,240) in cash donations from her business partners. Not everyone was happy about the ability of zone schools to acquire financial support from their business partners. A primary school head teacher reported receiving £400 ($640) from a local company to hold a Christmas party for his school. For him, this financial support represented "very little, miniscule amounts, £100 here, £100 there. Nothing. It's all sort of...mostly treats. Might buy the kids some sweets."

Most of the business support that schools in the zone received was in-kind. As one primary head teacher stated when asked about the type of corporate assistance she was receiving, "cash, no." The bulk of support from the North Upton's business partners involved employees from these companies volunteering as tutors within individual schools. The head of one primary school noted that the Bank of England "send in their volunteers who give up their lunch time, and they work with our more able readers and develop their skills." She went on to say that they are in the process of expanding this effort and have these volunteers tutor children in mathematics. A parent at another primary school noted that employees from J.P. Morgan volunteered to work as "reading" and "number" partners to tutor both gifted and underachieving students in these areas. The zone's director commented that this year his major voluntary sector partners paid for the cost of a full time teacher in one school and a second liaison worker in anther school as well as donating equipment worth £10,000 ($16,000). North Upton, like most EAZs, did not reach its targeted private sector contributions, but, according to the director, the combination of monetary and in-kind donations made the borough "one of the most successful sites" in the EAZ program.

One of the government's greatest concerns in introducing its Education Action Zone program was that Local Education Authorities would try to highjack this reform. As one member of the Department for Education and Employment's (DfEE) EAZ team, the government agency that initially administered this program, noted about the competition for zones, "the suspicion at the beginning was that LEAs would take over zones and there was a fear of that. The main reason that this came about was because we had two bidding rounds. In the first round, the LEAs had access to information, and they were able to write the bids... Most of the bids were drawn up by LEAs. The chairs at the forums were LEAs. The LEAs were quite big, and the fear was that this would be another LEA thing."

During the second round of bidding, in which North Upton was a successful applicant, the DfEE solicited proposals from other groups besides LEAs. This time, according to a member of the EAZ team, the local authorities "felt that they could get things going, and they did, and then they pulled back a bit. Now we see more genuine partnerships among those zones than in the beginning. The second round was different, [and] we saw more variety...The proportion of school led bids this time was half in comparison with about four percent the time before."

The North Upton application was submitted jointly by the LEA and a voluntary sector agency. On the face of it, the LEAs role appears to have been diminished. The administration for North Upton's zone was centered in two groups. There was the government mandated Action Forum that was chaired by a former head of a London financial trading company and whose members were zone stakeholders. They included two appointees from the Corporation of London—the city's governing unit, and representatives from the Labour government, the business community, the voluntary sector, the governing body of each zone school, teachers working in zone schools, and parents of children attending zone schools. The forum, however, did not have day-to-day operation authority of the zone. That power fell to a non-mandated executive committee that supervised the work of the Director and his staff and was composed of the Action Forum Chair, an appointee of the Corporation of London, a representative of a business partner, and three head teachers.[51]

Two researchers who have studied the EAZ initiative, Joe Hallgarten and Rob Watling, have argued, however, that the kind of joint bid submitted by North Upton and many other applicants during this second round was something of a ploy to disguise the LEAs role and assuage the government.[52] North Upton's application indicated that the zone, which would remain under the LEAs overall monitoring authority, would work collaboratively with other non-zone schools in the borough on an array of initiatives including professional development and school improvement.[53] A secondary school head thought that the LEA was playing a major role in the operation of the zone with a number of its staff working on school improvement issues with zone schools. The preference of the zone's director, who as a former secondary school head had worked with the private firm that was operating the North Upton's school improvement services, was to work through local authorities rather than private business. He found local authorities more accessible than this private firm and less driven by commercial concerns.

Although the role of the LEA may have been diminished in North Upton, and that is not exactly certain, the degree to which this governance structure represented an alternative to more traditional administratively dominated management schemes is less clear. The organizational structure was hierarchical with most of the authority concentrated in a small executive board that was dominated by zone administrators. This actually seemed to have been a common administrative pattern among EAZs throughout the country.[54] A member of the North Upton Teachers Association saw this concentration of power as being deliberate. "One of the complaints that I've heard about the EAZs is you have the schools, you have the Action Forum. You have the executive committee within the

Action Forum, and the whole process is one of excluding the people at the ground floor."

VIII.

Parents and community organizations, it seemed, played a small role in the management of the zone. The zone's director noted that they have virtually no parent participation and a few representatives from the community. One primary head noted that at least initially the zone was "very keen to get representatives from parent groups [and] very keen to get significant minority representation." During the first year that the zone was in operation, parent and community participation was high but by the second year their involvement had dropped significantly. One place where parents could have been represented was in the place reserved for members of the school governing body, many members of whom were parents. One head reported that at her school it came down to whether one of the governors or her would be on the forum. "The governors were very keen in the first instance to do it because they felt that somehow I might lead them up the garden path," she noted, "but in fact a governor who was elected and who was very keen to do it, I think, actually only managed because of pressures to get to about one meeting. So in fact, I am now back to being the representative." Similar situations may have occurred widely in North Upton since all of the head teachers that I interviewed reported being members of the Action Forum.

It is difficult to know why in fact parents did not participate in the governance of North Upton's EAZ. A primary school head teacher blamed the lack of parent involvement on the multicultural character of the borough. She thought that the ethnic diversity of North Upton made it difficult for any one parent or small group of parents to be selected as a forum representative. Zone administrators, however, had a dim view of the role that parents could play. The zone's director thought that most parents would be "bored stiff" if they were involved in the Action Forum. "What interests most parents in education is how their son or daughter is getting on in school. Now, I'm not going to tell them that; the forum isn't going to tell them that; the school's going to tell them that." He went on to say that his interest was in promoting the involvement of parents in the school, not in the administration of the zone.

Similarly, head teachers were not all that enthusiastic about parent involvement in the direction of the EAZ. One primary school head found parental involvement very stressful. Many of the demands that parents make, she noted, "are completely unrealistic. Some of the parental demands are very much focused on what they believe is best for their child with no notion that there are 200 other odd children in the school. I have difficulty with the notion that the tail is going to wag the dog. I'm very happy that parents work with us to improve the school, but I'm not quite so happy with the notion that parents are going to [run] the school."

Another primary school head felt that parent involvement in decisions affecting the zone created more problems than it solved. He argued that much of the opposition in North Upton to the establishment of the zone was the result of parents being manipulated and fed false information about the initiative by opponents who had harangued them at a series of meetings that were held to promote the EAZ in the community. "Parents were worried. And you know, when you're not involved in it, in education, and somebody says to you these project funds are coming in and they're going to take away your kids' pencils and things like that, you know... you're going to be worried." He went on to say that "parents didn't need to be consulted. You tell them. You say we're going in... you tell them after what the benefits are." He noted that in his school he consulted with the school governors, teachers, and because it was a Catholic school with the diocese and once they approved, he told the parents that the school was joining the zone.[55]

Parents, not surprisingly, did not feel encouraged to participate in governance. One parent and member of her school's governing body was active in the establishment of the zone, but once it was up and running she stopped attending Action Forum meetings. "If you expressed an opinion, you were in the way really. So really, the community dropped out." The forum, as she saw it, was primarily concerned about making "management decisions." She noted that "as a parent, you're quite stretched; you're trying to run your own life; your trying to do your own thing; it's quite hard to get empowered because you're busy doing things. It's easier for professionals to get up running." So in many ways parents think well, fine because they're so worn out."

This parent was not all that happy with her decision to withdraw from the Action Forum. Yet, constraints of time coupled with what she saw as a lack of encouragement from the zone's administrators led to her decision. "I think it's a shame. I think I would have carried on going to these meetings if I thought there was a way in which I could get involved, but I just didn't... I thought, well, they're getting on with it; they're doing their special stuff, but for me to take time out and get child care, and I have to pick up my kids and the meeting sometime is at 4. So it's impossible to juggle everything." Ultimately for this parent what was important was that the EAZ brought resources to the school that had "empowered my child... She was low achieving, and it's given her confidence." Having a say in the direction of the zone was not all that important to her.

IX.

As a reform initiative Education Action Zones looked quite different from the perspectives of its supporters and its critics. Its government supporters viewed EAZs as a reform that would remake Britain's system of state schooling. It would establish a new governance structure outside the LEAs that would open up the direction of education in England to partnerships involving parents, businesses, and voluntary organizations. The EAZ would provide for private sector financing of Britain's system of public education. The EAZ would allow its schools to depart from the National Curriculum as well as from the existing national pay scheme for teachers and provisions for their working conditions. It would also address the low academic achievement of urban youth, which in turn would improve their acquisition of job related knowledge and skills and enhance their employment opportunities. And over time the zone would be a vehicle for improving the economy of economically distressed communities like North Upton.

The way in which EAZ critics depicted this initiative as an effort to bootleg the Conservative Party's educational agenda under the guise of a left-of-center reform initiative reinforced the notion that this was a fundamental change in Britain's educational system. The emphasis they placed on certain features of the initiative—the proposals to alter the conditions of teachers' work and pay and to allow schools to depart from the National Curriculum, the threat of privatization posed by business partnerships, and the victim blaming view of parents and communities in New Labour rhetoric—went even further to suggest that this was a far more pervasive and ultimately more dangerous reform effort than the educational proposals of the previous Conservative government. Taken together, the supporters and opponents of EAZs created the kind of oppositional climate that pit unrealizable hopes against the dread of reactionary changes.

Those who led North Upton's Education Action Zone saw the program quite differently than did either its government proponents or outside critics. It appears that the claim made by EAZ critics that many of the program's goals were not achieved was largely correct. The North Upton EAZ remained under the control of its Local Education Authority. The governance structure remained hierarchical with authority largely in the hands of the LEA, school administrators, and the business partners. The influence that parents gained over the direction of the borough's education was minimal.

One of the greatest fears of those opposing the EAZ program was that zones would actually take advantage of the government's provision allowing them authority to depart from the National Curriculum and the pay and work conditions of teachers. The head teachers that were interviewed stated that neither of these policy changes was introduced in North Upton nor elsewhere. The National Union of Teachers (NUT) evaluation of the EAZ program in 2000 reported that no zones had made changes in the working conditions and wages of teachers.[56] The fact that nothing of this sort happened did not change the opinions that a group teachers from Camden, East London, and North Upton who I interviewed voiced about EAZs. From their vantage point, these were the very worst aspects of a bad program that thankfully did not occur.

One North Upton primary school head commented that "you hear this rhetoric that you can disapply parts of the National Curriculum. Oh yeah, then OFSTED will come in and they suddenly say you're not doing geography, you're not doing history. It's rubbish, absolute rubbish. I don't know of a single school that disapplies [the] National Curriculum." A secondary school head from the borough noted that they were able to offer a small number of students having special needs who were served at an off-campus site, a program outside the National Curriculum. They could not, she went on to say, do this for other students who were part of the school's general population.

Yet, the program in this borough did enjoy some successes. They were not necessarily the successes touted by the government. The great achievement of the EAZ from the vantage point of those in North Upton was that it provided them financial support, albeit limited by any absolute standard, to undertake an array of very typical school improvement efforts in ways in which they had become accustomed. With EAZ funds the schools of North Upton were able to support curriculum enrichment and remedial efforts; they were able to hire auxiliary personnel; they were able to improve facilities, particularly ICT; and they were able to provide additional professional development opportunities for teachers. The other important contribution that the EAZ program made to the schools of North Upton was that it brought in volunteers from the private sector who were able to provide supplemental instruction in such areas as reading and mathematics. In short, what the Education Action Zone initiative brought to North Upton was the ability to pursue traditional school improvement goals with some added resources. It was hardly revolutionary change!

There were certain features of the EAZ initiative that seemed to garner more support from North Upton's school leaders than others. The most popular aspects of the program were those that they could modify and adjust to fit their particular situation. What they liked best about the EAZ was that it offered them money to pursue goals and implement strategies and procedures that they thought were most appropriate for their students and their community. The features of the program that were least popular in this borough were those that threatened to disrupt accepted practices and those that compelled the schools and their staffs to follow procedures or work toward goals that were prescribed for them in advance and that did not lend themselves to modification. In that vein, teachers and administrators in North Upton were most opposed to requirements that they meet predetermined achievement targets, that they follow mandates to change the governance structure of the schools, or that they accept changes in their working conditions and salary.

The short life of North Upton's EAZ posed problems for its schools not unlike those encountered by schools supported by Annenberg money in New York City and Detroit. As we saw, these initiatives brought with them the financial and other resources that enabled schools to institute programs and services that seemed to bring about desirable improvements. The dilemma was that once these programs got going and were bringing about worthy outcomes, they were disbanded and the financial and other supports were cut off. School leaders and teachers were faced with the unenviable task of cutting back valuable programs or assuming fund raising tasks that detracted from their instructional responsibilities. Taking into account the significant problems of urban communities, this was an all too frequent outcome that undercut the effectiveness of schools as agents of change and betterment and did real damage to the lives of very vulnerable children caught within this flux of on-again, off-again reform.

X.

Finally, it is important to note that Education Action Zones were one phase of a larger strategy of New Labour to use the schools as an instrument of community building. As we saw the "third way" ideology subscribed to by the government had a particular understanding of community as a social entity comprised of individuals committed to a notion of common purpose and the common good built on the ideas of mutual responsibility and obligation. Within that context, New Labourites seemed to pay less attention to the structural problems underlying poverty and inequality and more on its roots in what they saw as individual pathology and family deficiency. Some of that thinking did find its way into North

Upton, at least in the proposal that was drafted to support the creation of the EAZ. Yet, it was muted, as we saw, by a recognition on the part of those who led the borough's EAZ that there were in fact structural causes to the North Upton's economic plights. In fact, once the EAZ was up and running, the focus of the program was solely on academic achievement. The issue of community building virtually disappeared.

The fact that those who led the EAZ initiative in North Upton had nothing to say about the issue of community in the day-to-day operation of the program does in itself tell us something about the success of the initiative as a mechanism of community building. One way to explore this issue is to consider the extent to which the EAZ possessed sufficient civic capacity to achieve its goals. There were early on some obvious warning signs that the EAZ program as a national reform lacked the kind of common understandings necessary to activate civic capacity. Most important in this respect was the program's short life. Taken together, the criticisms of EAZs raised by those outside of the New Labour government along with the quick willingness of the government to abandon the program suggested that its civic capacity was quite low. In the same vein, the gap between what the government sought in private financial support and what it actually received and the strong role that LEAs continued to play in the governance of EAZs suggested the inability of the parties involved in the EAZ program to secure the kind of broad agreement among cross sector partners that signaled the presence of civic capacity.

At first glance, we might conclude that events in North Upton pointed in the opposite direction. From the vantage point of those in North Upton, the EAZ had amassed important accomplishments. A key success of the zone, the director noted, was that "we galvanized and harnessed a great deal of support for schools from external sources." Because of this support, he went on to say, the zone was able to increase achievement "in specific areas," to enhance professional development, and to improve provisions for technology, which had been "in the dark ages." When he first became director he commented that he was "absolutely stunned by the lack of IT in both primary and secondary schools," but with the establishment of the EAZ, this became an area of strength. As a primary head teacher noted, this external support was helpful in "enriching the curriculum, definitely enriching the curriculum in a very major way, giving teachers a boost, giving them all these fantastic new resources...it was nice; it was just terrific."

Similarly, a secondary head saw the great success of North Upton's EAZ as its ability to establish linkages among the schools that allowed for cooperation and the establishment of a "shared vision of what the zone should be doing." There was, she went on to say, a clear focus in their work that remained constant but allowed for sufficient creativity to enable new ideas to be considered.

At the same time, however, there were indications that little in the way of civic capacity was activated in North Upton. The most striking evidence pointing to this absence was the opposition of borough teachers to the EAZ. Another such indicator was the fact that for many head teachers it was not so much the EAZ program itself that attracted them, but the funding it brought to their schools. My interviews also pointed to other possible signs of disagreement including the reluctance of some parents and school governors to join the zone, conflicts between school heads and their teachers over the zone, and the attitudes of head teachers regarding parental involvement in the management of the EAZ.

Looking further at the program's supposed achievements also raises questions about its actual civic capacity. The successes that those in North Upton attributed to the EAZs were, at it turns out, measured and modest. For the zone's director, the problem with this initiative was the array of expectations that the government had set for it. It was, he commented, "pure nonsense" to believe that one initiative operating on less than £1 million could transform everything. The government, he went on to say, "wanted every indicator to go up on a million quid a year," and one "couldn't work intelligently on that basis." What an EAZ had to do was to select a few "key areas" on which "to move forward." He believed if you did that and then had demonstrated accomplishments, the enterprise was successful. And that was precisely, to his way of thinking, what was occurring at North Upton.

The problem that this initiative posed, according to one primary head, was that it built up expectations, promoted a number of good initiatives, and "then suddenly it's all gone." As he saw it, once the money came to an end, the program was finished. "Money is tight. And I am sorry, it's been wonderful, but we will not be able to afford emotional-behavioral counselors, the French, the sports—we will not be able to afford that." A secondary school head noted that the demise of the EAZ will be "devastating" to her school. Because of the EAZ, she was able to provide parent workshops, programs for disaffected students, and cooperative ventures with other schools. A primary school head noted that when the EAZ was established, North Upton citizens were told that "it's for the long term . . . and then the money is taken away. Well it is a fad. It was a passing experiment, but because New Labour didn't hit their targets, their obsessionally bloody targets—you know, they suddenly pulled the funding." He went on to say that he did not understand the government's preoccupation with targets. He pointed out that last year their school did well in English but terrible in mathematics. This year, however, many of his students were doing better in mathematics. New Labour's targets, he felt, were "meaningless." It would be better, he thought, if everyone would simply say that we "will all do our best."

It was the impending demise of the EAZ that was most troubling to those in North Upton. They probably could continue to fund some of the projects in their schools. The director pointed out in this vein that if a school wants "to keep attendance high, they'll carry on purchasing a home-school worker out of their own money." The schools within the zone, he argued, would have to decide what their priorities were. "When you've had funding for four years and you know its going to end at the end of year five, well then you have to make plans accordingly. I've got no sympathy for the school that says well, you know, what are you going to do? Well, you've been lucky so far; you should have been…prioritizing for the future."

XI.

In this chapter we have continued our discussion of partnerships by looking at New Labour's Education Action Zone initiative. What we discovered here was not all that different from what we found in our exploration of the Annenberg funded efforts in New York City and Detroit and the Youth Trust venture in Minneapolis. Partnerships enabled schools in all of these settings to collaborate with a variety of external stakeholders, who in turn could provide financial and material resources as well as expertise to enhance their educational programs. In all of these settings, partnerships were a source for funding and supporting personnel and programs that these schools had heretofore been unable to provide.

The rationale for entering into these partnerships in both the United States and Great Britain was that the resources that these collaborations provided would enhance student academic achievement. Here the picture was at best mixed. There were indictors in New York City's Annenberg Challenge that student achievement did increase. Yet in Detroit, indicators of achievement gains were uneven. The results from the EAZ initiative was also inconclusive. Where supporters of the venture pointed to a number of indicators of improved student performance, its critics challenged those reports. This was a pattern of uneven performance that paralleled the impact of this reform in North Upton.

An important goal of these partnerships was to introduce major organizational and administrative changes into existing state systems of schooling. Such changes never occurred either in the United States or in

Britain. Efforts in both settings to use partnerships to create new organizational schemes did not succeed. Schools participating in these partnerships looked pretty much the same when these ventures were inaugurated and when they came to an end. In this vein, the partnerships' stakeholders had very different views concerning the goals of these initiatives. Promoters of partnerships seemed to embark on these ventures in hopes of changing to varying degrees the governance of public schools. This was probably more so the case with the supporters of Edison's charter school venture in New York City and New Labour's EAZ initiative than it was with those involved with the Annenberg Challenge or with Youth Trust. It was the fear of such fundamental changes that led critics to oppose partnership initiatives, particularly those that involved business. In this vein, however, it is important to note the support that teachers and administrators in North Upton offered the EAZ initiative was not so much because of the prospect it offered for change but rather for the financial resources it offered them for pursuing traditional and well accepted school improvement efforts.

In both the United States and Great Britain the reliance of school reformers on partnerships spurred opposition. The concern of those who criticized such arrangements was that they would provide an opening whereby business interests would be able to exert influence and ultimately control over public education. Partnerships were, to their way of thinking, vehicles for privatization, which in turn would undermine the democratic goals of public school in favor of commercial and corporate purposes.

As events turned out, no such business control ever occurred in the partnership ventures that we considered. Probably the worst that can be said about the actual role of business in these partnerships was that they may have directed participating schools in directions that were unlikely to be sustainable or at least difficult to maintain once the partnership came to an end. On the other hand, it may be the case that school-business partnerships could serve as instruments for bringing together different stakeholders with distinct interests around the common purpose of school reform.

Our discussion of the Education Action Zone initiative noted that to the extent that partnerships reflected "third way" thinking, they were embedded with a very particular ideology about the role of education in addressing the problems of inequality, especially in urban communities. Proponents of EAZs downplayed the structural explanations rooted in poverty, racism, and unemployment that are often given for the unequal outcomes that plague economically distressed schools and their clientele. Instead, they attributed low academic achievement and other similar problems to deficit families and communities and saw the task of school reform as that of instilling the poor and disadvantaged with the value of responsibility and obligation that would lead them to acquire the knowledge and skills that they needed for employment. The society that these "third-way" thinkers envisioned was one that might provide at some level support for those in most need. It would not, however, do so by equalizing in any way the conditions that divided individuals from each other and groups from groups. Rather, it would do so by providing a minimal social safety net for those most vulnerable and exact in return a commitment to life long learning and employment. It is not necessarily the case that all partnership ventures must be rooted in the same ideological assumptions that grounded Education Action Zones. Yet, as vehicles of educational reform, there is the danger that partnerships can accentuate the tendencies often attributed to globalization of reinforcing inequality and social stratification.[57]

Finally, we need to ask to what extent the idea of community has provided us with a lens for understanding the role of partnerships as an instrument for urban school reform. Community, as it turns out, has not only been my conceptual framework throughout this volume. In the case of the EAZ initiative, the idea of building community was an explicit notion that proponents of this reform invoked in their advocacy of partnerships. EAZs, then, offer a good opportunity to consider how an avowed commitment to building community actually played itself out in practice.

The definition of community that we have employed in this volume pointed to the activation of civic capacity as one good measure of the existence of community. From that vantage point, it seemed that the EAZ initiative was so mired with conflicting interpretations that a search for community was fruitless. Looking more broadly at the full range of partnerships that we considered in these last two chapters, it would seem that the best that can be said about the community building potential of this reform was mixed. There were indicators within these reforms that civic capacity was activated and a sense of collective belonging was achieved. There were, however, other indicators that these results had not occurred. Yet, using the notion of community as a conceptual framework did have a particular advantage in that it led us to consider the ideological assumptions guiding partnerships. In identifying a "third way" perspective as offering that grounding, our examination has left us with some pause as to the viability of partnerships as a vehicle for urban school reform.

Thinking about urban school reform from the vantage point of a notion of community leaves us with the question of how proponents actually go about changing an institution. This is a complex effort that involves an array of tasks including but not limited to developing a vision for change and establishing relevant goals, securing stakeholder buy-in, exercising leadership, and addressing obstacles. These are the issues that we will address in our next and last chapter as we employ a concept of community to explore an actual instance of school change.

Chapter 7

Smaller Learning Communities and the Reorganization of the Comprehensive High School

Barry M. Franklin and Richard Nye

I began this volume by looking at some of the findings that my research assistant and I had uncovered in our study of the establishment of smaller learning communities at the school I am calling Timberton Central High School. What we found and what led me to write this book was the role that a notion of community seemed to play in defining what this reform was all about and consequently my belief that this notion could prove useful in interpreting other past and present urban school reform initiatives. The purpose of this final chapter is to look further at Timberton's effort to reorganize itself around a number of smaller learning communities and to see what a notion of community tells us about that endeavor.

Timberton is a medium sized, industrial city with a population in 2001 of around 79,000 inhabitants located about 40 miles north of the state's capitol city.[1] In that year, Timberton Central High School embarked on an effort to convert this traditional comprehensive high school into a number of smaller learning communities. The impetus for this initiative was a demographic shift in enrollment that was bringing into the school increasing numbers of at-risk, limited English proficiency students. This was a population change that was the result of a post–World War II in-migration into the city of an economically distressed, low income, largely Latino/a population. For several years now, Timberton has had the highest concentration of Latino/as of any community in the state. Between 1990 and

2000, their share of the city's population increased by approximately 138 percent.

Exacerbating this transition was Timberton's own economic decline. A largely residential city with a diverse industrial base made up of sugar processing plants, flour mills, grain elevators, meat packing plants, livestock feeding yards, freight terminals, and aircraft industries, the city underwent a number of economic disruptions in the decades following World War II. The most important of these was the plant closings, subsequent job losses, and eroding tax base brought about by the demise of the city as a passenger rail hub. Also important was the displacement of farming and agricultural related industries by residential growth in the surrounding county, the deterioration of the downtown commercial sector and the relocation of its businesses to surrounding suburbs, and the instability and fluctuating health of the defense related industries that dominated the city's manufacturing base.

The result was high levels of unemployment among the city's inhabitants, a growing population of unskilled individuals working in low paying jobs, and a high rate of childhood poverty. Its low median family income of $38,950 in 2000 was 22 percent lower than the national median. Together with its high rate of childhood poverty and its concentration of ethnic minorities, Timberton qualifies as the states only designated federal Enterprise Zone.[2] According to Timberton Central's principal, the city's "industrial base has "essentially evaporated. It was a powerhouse in the '50s and '60s, and that has long since died. So tax base wise [Timberton] is hurting." This was an outcome not dissimilar to what we saw in chapter three in our discussion of the impact of deindustrialization on Detroit. Although Timberton remains a railroad hub for freight traffic, is an important tourist destination for summer and winter outdoor recreation, and is enjoying some success in attracting high tech manufacturing companies, it is a city with significant economic problems.

This demographic reality has not left the city's schools unaffected. Timberton Central has seen significant growth in its Latino/a student population. In 1999 there were 282 Latino/a students enrolled in the school. Two years later, the number had risen to 486. Today, approximately 43 percent of the student population is of Latino/a origin and 23 percent possess limited English proficiency or speak no English. The school has also seen growth in its numbers of economically disadvantage students with over half of its enrollees qualifying for free and reduced lunch. Accompanying these change in the student population have been declines in student academic achievement and graduation rates and increases in drop out rates, student drug use, and school related violence. In 2000–2001 and in 2001–2002, Timberton Central failed to make Annual Yearly

Progress (AYP) as measured by the state criterion referenced tests. The 2001–2002 tests indicated that 65 percent of the school's students scored below the state standard in science, 38 percent below the standard in language arts, and 80 percent below the standard in mathematics. And in the same year, Timberton Central reported graduation rates of 68 percent for its White students and 32 percent for its minority students.[3]

II.

The appointment of a new principal at Timberton Central in 2000 was the starting point for the school's reorganization into smaller learning communities. A former assistant principal at a high school in a nearby district, Timberton's new leader noted that when he was interviewed for the position, he was asked how he would change the school. He responded by indicating that he would approach the task much as one would do in preparing for an external accreditation. As he noted, "I would take a look at the data, and based on the data, I would develop some action steps. And that is precisely what we did." During his first year, he went on to say, the school undertook a self study during which "we just looked at what we were, and we weren't very good, and we weren't getting any better." The assistant principal echoed the same viewpoint. "We realized that we weren't very good, and we weren't going to get any better doing what we were doing; we weren't going to get better kids; we were going to continue to get the kids with the demographics that we had. And that if we're going to improve achievement, we needed to do something ourselves."

Timberton's self study was part of a district wide effort involving both of the city's high schools in addressing the problem of low student achievement. Timberton Central and its sister school established school improvement teams comprised of administrators, teachers, students, parents, and community leaders to study the prospects for reform and submitted a number of state and federal grants to support this work. A requirements for securing one of these grants, a state sponsored Comprehensive School Reform Demonstration Program grant, was for the schools to identify and align their reform initiative with one of a number of recognized comprehensive school reform models.[4] For his part, Timberton Central's principal noted that he began to look at the existing effective schools research, attended an effective schools conference, and visited a couple of schools in Delaware whose involvement with the High Schools That Work initiative "were really impressive and kind of excited us."

High Schools That Work is a school improvement program established in 1987 by the Southern Regional Education Board (SREB), an interstate compact comprised of sixteen Southern states. Made available to schools outside of its region, the initiative is designed to introduce and support a college preparatory curriculum comprised of academic and career/technical studies that its promoters claim cultivate high standards and enhance academic achievement.[5] It was, according to Timberton's principal, "the only effective high school reform model that has consistently shown results, and so that's how we became involved and engaged and married to those people."

High Schools That Worked offered Timberton a recognized model of comprehensive school reform that supported a number of practices that seemed to direct Timberton toward its goal of building a school culture of high academic achievement. Among them were integrating high expectations into classrooms, offering a curriculum including academic and career/technical studies, providing for work-based learning, supporting teacher collaboration, and promoting active student engagement.[6] Smaller learning communities, however, were not part of the High School That Works agenda. As the school's principal noted SREB was not a supporter of smaller learning communities and "at some times they actually were working counter to us." He went on to say that some of the practices called for in the initiative "were irrelevant to our work in learning communities, and so as we worked with them, there were some tensions that developed over time... So it wasn't the best marriage but it was useful to us."

The impetus for embracing smaller learning communities came from a number of places. The schools that Timberton's principal visited in Delaware and so impressed him had been organized into a number of smaller units or academies as they called them. A second source for the reform were criticisms of large comprehensive high schools coupled with the support for smaller learning communities that were prominent in the research literature that Timberton's staff encountered in preparing grants to support their reform efforts.[7]

A third reason behind the interest in smaller learning communities may have been the district's experience with a similar organizational pattern. Over a decade previously, according to a central office administrator who was responsible for developing grants to support the smaller learning communities venture, Timberton Central's sister school had introduced a freshman school-within-a-school that reserved a wing of the school for a separate program for entering ninth graders. An English teacher at the time, she notes that this reorganization allowed for a more flexible class schedule, team teaching across subject areas, the use of interdisciplinary projects, and service learning. "I could," she pointed out, "have the kids all day [or] I could take the kids for an hour and a half on Monday because I am doing a project." She went on to say that she had recently spoken to students who had gone through the program and asked them about their feelings of connection to the school and "they all said they felt connected more in the ninth grade "because we had the same teachers and because we had the same kids. We felt like we were a community... The teachers called on us. They knew us. We knew all of the other kids, and we felt safe."

Once having selected smaller learning communities as their model and securing some grant money, the principal sent groups of teachers to visit small schools and smaller learning communities around the country to get ideas for their reorganization. One of the difficulties that staff members noted was in finding programs that were sufficiently similar to Timberton to guide their work. Several teachers traveled to New York City to visit a number of small autonomous high schools housed in the Julia Richmond Education Complex. An English teacher who was part of the group commented, "I realized that you know we are different because they have smaller schools... and not smaller communities, and so schools of 300 are different than a community of 300. He went on to say that at Timberton, students identified with their school, but he did not know if that would transfer to identifying with a community. He noted that when New York City introduced small schools they shut down some of their large comprehensive high schools and renovated the buildings. Students attended several small schools and shared the same building but "they don't consider themselves as going to the same school. They have different names... "

Another English teacher who visited Julia Richmond noted that "their concept was very different from what our concept is... They took a large building and divided it up, but what they did was... [take] all of the kids out of there first and then brought them back. Then reconfigured and brought them back. And we don't have those options that they have." These schools, she went on to say, were not truly comprehensive high schools. She did not see any of the computer labs or sports programs that existed at Timberton Central. An ESL teacher pointed out that every school that they visited was different, and it was unclear to her how they could implement this or that practice at Timberton. She came away from the visits noting that "there is not a right or wrong way of doing this."

A child development teacher who would ultimately lead Timberton's Health Science and Human Services Community made clear how different many of the schools that were visited were from her school. "There was no way to shut the school down and ship the kids elsewhere like some places in San Diego have done and totally restructure the whole school and hire new teachers, and then bring the kids back in as part of a small school.

We don't have the staff to do a small school. I know. I'm thinking of San Diego Unified because I just went to a conference this summer on what they have done. You know they have enough staff where they can isolate each teacher into the different schools, even if they are on the same campus. So like if I were a family and consumer science teacher—most of our programs are in heath science and human services—so I would teach only to those kids and some kid who is in applied science and technology wouldn't be able to take child development because at the conference they talked about they reserve a few spots in certain areas for kids outside the different schools,...but they are totally autonomous schools. And that would kill our program." A math teacher also noted the differences between what they saw and Timberton. "There was more autonomy within each school than there is [here]. I mean that they had their own grading. They had their own principal; they had their own area of the school...And so we are not able to do that yet."

Nonetheless those making these visits thought that they could learn from the experiences of these other schools. A social studies teacher and coach pointed out that staff members visited a number of schools but didn't think that they had found one that was like theirs, particularly with respect to the ethnic diversity of the student population. He thought that they were able to get some good ideas that they were able to use in their setting. A math teacher felt similarly. As she put it, "we could pool or pull together and share the good points from our observations that would meet the needs of the student here. So we could pluck and pick and get all of these ideas—diverse ideas from various learning communities—and pick out the best..."

III.

If we are to understand the role that smaller learning communities are designed to play at Timberton Central, we need at the outset to situate that initiative in the historic debate that educators have had concerning school size. A 1959 U.S. government statistical report noted that from the first "smallness" has been a pervasive feature of the nation's public schools.[8] This has certainly been the case with the high school. The period between the last decade of the nineteenth century and the first two decades of the twentieth century saw a tremendous expansion of American secondary schools from around 2,700 to almost 20,000 and an enrollment growth from around 211,000 in 1890 to over 2.5 million in 1924. By 1958, high school enrollments reached seven million students or somewhere between

70 to 80 percent of those between the ages of fourteen and seventeen. Yet, these high schools were relatively small institutions. There were certainly large schools with enrollments over 500 and even a few that housed more than 5,000 students, but the median size of the American high school in the 1923–1924 academic year was around seventy-seven students.[9] Over time, high schools would become larger, but at the end of the decade of the 1950s, more than half of the nation's high schools enrolled less than 300 students.[10]

School size, as it turns out, was viewed differently during the first half of the twentieth century than it is today. As we have suggested already and will expand upon later in this chapter, the small high school has come to be seen as an instrument for attaining a number of desirable student outcomes, not the least of which is improved academic achievement. Earlier in the century, however, the small high school was seen as something of a problematic institution that did not provide its students with an adequate education. There have been over time some differences of opinion as to what constituted a small high school. Writing in 1926, John Rufi, an education professor at Michigan State College of Agriculture and Applied Science, identified small high schools as those that enroll seventy-five or fewer students and employ four or fewer teachers. A 1937 study of California's small high schools set 350 students as constituting the maximum enrollment of what its authors referred to as the small school. For James B. Conant in his famous 1959 study of the American high school, the small high school was one whose graduating class was less than one hundred students. And a few years later, a National Education Association report labeled high schools with fewer than three hundred students as small schools.[11]

From the standpoint of professional educators, small high schools posed a number of problems. Among those Rufi noted in the mid-1920s were the lack of well trained and experienced teachers, inefficiencies in administration, high operating costs, an inadequate physical plant, a lack of extracurricular activities, and a limited curriculum. Twenty years later, University of Southern California professor E.H. LaFranchi noted a similar set of problems.[12] Among these difficulties, the curriculum stood out as key. Small schools, so the argument went, often did not have the personnel and other resources to offer more than a single, college preparatory course of study. They could not provide the array of non-academic programs, particularly vocational courses, to meet the needs of all of those who might wish to attend.[13]

It was this criticism of the curriculum that existed in many small schools that formed the basis for James B. Conant's support in his study of the comprehensive high school during the late 1950s for efforts at reducing their number through school consolidation. Echoing these earlier critics of small high schools, Conant argued that schools with a graduating class of less than one hundred, the defining feature of the small school to his way of thinking, could not provide an adequate education to any group of students including those who were academically able, intellectually slow, or vocationally oriented. The typical curricular option available to students attending these small schools, he noted, was an academic one. Requiring all students to enroll in such a course of study had the effect of frustrating less able students while reducing standards for those who were more able. Replacing these small high schools with larger comprehensive ones would, he argued, allow for a system of secondary education that would meet the needs of all adolescents.[14]

IV.

Why is it that an institution that was viewed for most of the twentieth century as problematic is currently one of the most popular approaches for reforming the high schools? The answer in part lies with the fact that school reform is not a monolithic enterprise. Rather it comprises a number of circulating discourses—in the form of beliefs, knowledge, and social practices—whose movement across time and space define and assess small school reform in different ways.[15] One such discourse is represented by the criticisms of small high schools that we have just considered. There are, however, other discourses that form further lines of descent that shape today's small school reform movement. There is a discourse that looks at small schools in a more positive light. Writing in 1961, Paul Smith reported on a study that he had conducted a few years earlier on the impact that high school size had on the academic performance of first year college students. While students who had attended the largest schools had higher scores on tests of language, reading, mathematics, and overall academic achievement than students attending smaller schools, the differences turned out not to be statistically significant. As Smith saw it, both small and large high schools could provide students with a quality education and that in the end it was not the size of the school that mattered but its ability to teach its graduates to read, write, and think.[16]

In a 1964 book, Roger Barker and Paul Gump also questioned the claims about the inadequacy of the small high school. Their study examined differences between small and large schools in what they called behavior settings—their term for an array of features of school life including classes, athletic events, physical sites within the schools, and student club meetings— where student carried on their "school lives."[17] Looking at thirteen high schools in Eastern Kansas that ranged in enrollment from 35 to 2,287 students, they found that small schools were not all that different from large schools in what they could provide their students. Small schools, they argued, were able to offer their students many of the same services and support many of the same events that were to be found in larger schools.[18] In the end they concluded that there was a "negative relationship" between the size of the school and what it offered in the way of "student participation." As they put it when "schools become more heavily populated, more of the students are less needed; they become superfluous, redundant." What was desirable, Barker and Gump concluded, was a school in which students should not become "redundant."[19] Three years later the superintendent of the Julian Union High School District in California noted that the introduction of modular scheduling would enable a small high school to offer a greater number and variety of courses than many larger schools while providing students with such desirable features of small schools as increased personal attention and greater access to extra-curricular activities.[20]

Even those who offered criticisms of small high schools noted their strengths. The author of the 1959 statistical study of changing school size that we cited earlier noted the difficulty that small schools had in meeting the myriad demands placed upon them. Yet, he pointed out that in many of the sparsely settled rural communities of the nation, the existence of small schools was unavoidable. He went on to say, however, that such schools had certain desirable features:

> Moreover small local staffs continue to have the advantage of knowing the children, their homes, and their parents intimately enough to provide a personal, family-like situation in the school. Small schools tend to be near the farm homes, to engender parental confidence, and to serve community functions of recognized value. The small schools also continue to play an important role in local self-government.[21]

A few years later a critic of small high schools noted that despite claims about the deficiencies of such schools, there were situations when it was better to keep these schools rather than transporting students a great distance to larger schools. Small high schools, he pointed out, could if they were well equipped provide as good an education as larger schools. Beyond that, however, he noted that small schools "keep the function of education in the community, help maintain a community center within easy reach of every farm and village, and because the classes are small, are more conducive to quality education."[22]

Giving some impetus to this positive view of small schools was a discourse that emerged out of the ongoing debate among twentieth century educators over the organization of the curriculum. On one side stood those who favored organizing the school's program around the traditional disciplines of knowledge. Pitted against them were a loose collection of individuals who proposed other organizational patterns derived from the interests and concerns of youth and the social problems affecting the larger society.[23] Among this latter group were those who looked to the ideas of educational progressivism and particularly the work of John Dewey for many of their underlying beliefs. Of particular importance in this vein was an understanding of children as active constructors of their own learning, a conception of the school as a model of a democratic and just community, and such curricular and pedagogical practices as curriculum integration, project based learning, the role of the teacher as facilitator, and a course of study that connects the school with the surrounding community.[24] It was a viewpoint about curriculum and pedagogy that would provide support for the movement for the establishment of small schools.

Paralleling the commentaries that we have considered about school size was a related discourse about class size that would also help shape today's small school movement. Although Conant advocated the establishment of larger schools, he was mindful of the dangers that were posed by the resulting enrollment growth within the comprehensive high school. He recommended in this vein that the pupil load for teachers of English composition be limited to 100 students, down from the 180 he saw in some of the schools that he visited for his study.[25] In making this suggestion, he was entering into the longstanding debate among American educators about a question that was clearly related to school size, namely that of class size.

Class size has been an issue that American educators have wrestled with in one form or another since the introduction from Europe in the early nineteenth century of the Lancastrian or monitorial system. On one side of the debate were the proponents of this approach with its use of student monitors under the direction of a teacher to instruct large groups of children, often several hundred at a time. Pitted against them were supporters of teacher led patterns of simultaneous recitations with smaller groups of children that ultimately became the graded classrooms that we are familiar with today. This latter method of instruction still resulted in teachers being responsible for large groups of students, certainly over the twenty-five or thirty-five we see in contemporary classrooms, but nowhere near the hundreds of students provided for in the monitorial system.[26]

Research on class size at the secondary level over the course of the twentieth century has yielded inconclusive results. A 1928 report on class size and efficiency in San Francisco's high schools noted that there was no way of "definitely determining this point." The committee that prepared this report noted that there were teachers in the city's high schools who could teach large classes and other who could not. The report cited studies that both identified the most desirable class size for efficient instruction and those that indicated that these factors were unrelated.[27]

A summary of a number of studies, largely in elementary schools but also including high schools and universities, undertaken during the first three decades of the twentieth century reported by Milton and Dortha Jensen in 1930 pointed in contradictory directions. They noted that the studies in elementary schools that used promotion rates as the criterion of student success indicated that those enrolled in small (less than twenty students) and medium classes (twenty to thirty students) performed somewhat better than students in large classes (thirty to fifty students). Yet, they also cited a study using standardized test scores as the measure of student achievement that found that elementary students in classes from forty-five to fifty-five perform as well as students in smaller classes.

At the secondary level they cited a North Central Association study using grades as the criterion of student achievement that reported differing results according to the subject investigated. Students enrolled in small classes in English earned higher grades than those in medium or large classes. They found, however, that students enrolled in large classes in social sciences and mathematics received higher grades than students in small and medium sized classes and that those in medium sized modern language courses earned better grades than students placed in large modern language classes.[28] In a follow-up article that Milton Jensen published several months later, however, he reported that his own study in second semester algebra courses, conducted under more controlled conditions than was the research that was cited in the earlier article, pointed to higher levels of achievement for students enrolled in small classes.[29] Subsequent research on high schools has done little to resolve this issue of class size. There have been studies that point to higher student achievement in small classes as well as research that claims that that greater student success occurs in larger classes. And there have been studies indicating that class size makes no difference in student achievement.[30]

A study that began in Minneapolis in 1943 produced inconclusive results concerning the value of smaller class size. In that year Newton Hegel, Principal of Minneapolis's Folwell Junior High School introduced what came to be called the B Curriculum or Small Class Experiment. Designed for students who were thought likely to receive failing grades in the school's regular classes, which enrolled around forty students, this initiative provided these students with classes in English, social studies, and mathematics that were half the size. As Hegel saw it, reducing class size by itself would be sufficient to ensure the success of these potentially failing students. His viewpoint was, however, controversial and with the expansion of the program to all of the city's junior high schools in 1946 differences emerged among teachers concerning how these classes were to operate. Some teachers followed Hegel's lead and except for class size maintained the regular curriculum. In some schools, however, changes in curriculum and evaluation were introduced that had the effect of turning these smaller classes into remedial programs. A 1947 citywide assessment of different approaches that the city was then offering in its junior high schools to address academic failure produced varying results. Some schools offering the B Curriculum saw improvement in student achievement, while others did not. As a consequence during the next two years, these small classes were disbanded throughout the city.[31]

More recent research on this subject, which has focused its attention on children in grades kindergarten to three, sees smaller class size as making a positive contribution to an array of desirable outcomes including student academic achievement.[32] The most well known of these research studies, Tennessee's Star Project that was conducted between 1985 and 1990, reported that students in classes of thirteen to seventeen in kindergarten through grade three outperformed students in regular classes with enrollments from twenty-two to twenty-five as measured by standardized tests in reading and mathematics. In addition, follow-up studies found that children enrolled in the small classes were more likely to be promoted from grade to grade than students in regular size classes, were more likely to take advanced courses in high school, were more likely to graduate high school, and were more likely to attend college.[33]

It was the ebb and flow of these discourses over time that began to shift opinion away from earlier criticisms toward a more favorable climate for efforts to restructure high schools into smaller institutions.[34] A 1972 paper prepared for the San Francisco Board of Education on future directions for the city's secondary schools noted a range of problems facing the district including high drop out and truancy rates, increased crime, low student achievement, an out of date curriculum, and schools that were too large and impersonal. The paper went on to detail a number of ways to address these problems including a reorganization of the high schools into smaller units.

One proposal called for the transformation of the city's comprehensive high schools that typically enrolled anywhere from 2,400 to 3,100 students into schools-within-schools with enrollments limited to 1,500 students or less Under this plan, the ten academic departments that existed within a typical city high school would be reorganized into two houses of 750 students each. The houses themselves would be subdivided into modules of 250 students in which interdisciplinary teams of teachers would provide instruction in humanities, science, and mathematics. Other subjects, including physical education and music would be taught in more traditional classroom settings that drew their enrollment from both houses. The proposal also sought to remedy what was viewed as the impersonal character of comprehensive high schools by proposing that students would remain the in the same module during the three or four years that they attended the school. Doing so, the paper suggested, would ensure that students would be able to establish a personal relationship with several teachers during their high school career.[35]

These first initiatives to transform high schools into smaller units during the 1960s involved the establishment of small alternative schools, typically located in large cities, for students who were not succeeding academically or behaviorally in traditional high schools. The next four decades would see an expansion of these efforts fueled by private foundation and government grants to create small schools and smaller units within larger schools to serve not only at-risk students but those preparing for higher education as well.[36]

What distinguished these schools from traditional comprehensive high schools was their embrace of specialized curricular and pedagogical approaches. Some, such as Central Park East Secondary School that Deborah Meier led in New York City during the 1980s, were organized around a child-centered and constructivist instructional model rooted in the principles of educational progressivism.[37] Others, such as the over one hundred small schools that were developed during the 1990s as part of the Annenberg Foundation funded New York Networks for School Renewal, adopted an array of guiding themes. There were schools within the Networks that integrated academic programs with the arts, schools organized around such subjects as science and technology, and schools that promoted community service, global awareness, and social justice.[38]

A good example of the diversity of themes around which small schools are organized can be seen at California's Berkeley High School. During the course of the 1990s the school was reorganized, although it maintained its identity as the city's single comprehensive high school, into six smaller units. One part of the school was comprised of four relatively autonomous and self contained small schools—Arts and Humanities, Communication Arts and Sciences, Community Partnerships, and Social Justice and Ecology—that offered their own specialized curricula reflecting their different organizing themes. Students took the majority of their courses from the specialized offerings of these schools supplemented by courses necessary to meet college admission requirements offered by the school's comprehensive departments—African American studies, Visual and Performing

Arts, Mathematics, Physical Education, Science, Technology, and World Languages. The other part of the school was comprised of two less autonomous and less self contained smaller learning communities—Academic Choice and the Berkeley International High School—which offered college preparatory courses of study, one rooted in the humanities and the other meeting the requirements for the International Baccalaureate. Students in these programs took two or three courses a term within their community and the remainder of their courses from the offerings of the school's comprehensive departments.[39]

There were, as it turned out, a number of dilemmas that faced proponents of small high schools when it came to the actual task of transforming a school district comprised of one or more large comprehensive high schools into smaller ones. One issue had to do with how the conversion would occur. Would buildings that housed single comprehensive high schools be physically altered to accommodate a number of smaller schools? And if this were the case what would happen to the students during the conversion, particularly if the redesign required the temporary closing of the building in question? Where would these students be housed? Would the district have at its disposal available physical facilities for this to take place? What would happen if this redesign was to occur in a small district with only one high school? If this conversion resulted in the establishment of new schools would the district have the teachers and administrators to staff these new facilities. Finally, where would the money come from to support the various costs associated with this conversion?

A possible solution to these problems that some supporters of small schools have put forth is the concept of smaller learning communities.[40] The most recent version of what has been referred to as academies, house plans, or schools-within-schools, it is not always all that clear how this approach differs from that of a small school. As we saw in our example of Berkeley High School, there did not seem to be that much difference between these two methods of downsizing that particular institution. Yet there are some common features of smaller learning communities that often distinguish them from small schools. Smaller learning communities are units within larger high schools and consequently require less in the way of physical changes to existing buildings. Such changes can be undertaken, but they do not have to occur. A certain portion of an existing building or a floor of a multi-storied building can be designated as the smaller learning community without requiring any physical renovation.

Smaller learning communities typically have less in the way of autonomy than small schools. While small schools have their own administrators and teaching staffs, a number of smaller learning communities within a school may share their personnel. Where small schools can be independent entities within a district under the direct supervision of a school system's central administration, smaller learning communities are more likely to be part of an existing comprehensive high school and under the direction of that school's building principal. Finally, there can be significant curricular differences between small schools and smaller learning communities. Small schools typically have their own, unique specialized curricula and operate as self-contained entities. There are unique and specialized features to the curricula that smaller learning communities housed within a comprehensive high school offer. Yet, these smaller learning communities rely to varying degrees on a common curriculum that is shared by a number of such communities occupying a single building.[41] For Timberton Central High School, it was the smaller learning community that served as the model in its reorganization effort.

V.

My research assistant and I arrived at Timberton Central at the end of the summer of 2006. At that point, the school was beginning the fourth year of its reorganization into smaller learning communities. It had in the previous year established its ninth grade center and its four communities, assigned teachers and students to their respective units, and was in the midst at different stages within each of the communities of developing and implementing its theme based curriculum. During the course of the year we visited the school, sometimes together and sometimes singly, approximately every two weeks for two to three hours at a time. As part of our time in the school, we interviewed teachers, students, and administrators, talked informally to teachers in their lounge, sat in the hallway outside the main office and observed passing events, attended learning communities and school wide meetings, and toward the end of the year observed classes.

In the course of our year in the school, we collected a large body of observational and written information. The most important data sources for our study were the interviews that we conducted in our visits to Timberton. Supplementing these data were field notes that each of us took during and after our visits to the school and archival records in the form of teacher class schedules, meeting agendas, memorandum of various sorts, and teacher produced curriculum materials. In reading the written transcriptions of our interviews and focus groups we found that those to whom we talked addressed seven key issues. We have already explored the reactions of those who visited other schools in preparing the school's reorganization. Other issues that were discussed included the leadership that administrators and others exhibited during the planning and initial implementation of the initiative; the extent to which teachers and others supported the school's reorganization; the objectives to which the initiative was directed; the effect of smaller learning communities on interpersonal relationships among teachers and students; obstacles that slowed down the implementation of the project, and conclusions about the accomplishments of the project.

As was mentioned in section II, the Timberton School District searched for and found a principal who was capable and committed to instituting a change process. The principal for Timberton Central began writing for grant money to overhaul the school after consulting literature that detailed various reform models. Support for his effort was recognized as Timberton Central received nearly one million dollars over the course of three years in federal funds. The district also contributed nearly $400,000 in matching funds for the smaller learning community initiative.

The faculty quickly recognized the principal's commitment to smaller learning communities as he shared with them his research findings. A social studies teacher in the Ninth Grade Community suggested that she was on board with what the new principal wanted to do because of his "research savvy" and "intelligence." The Timberton District office quickly supported their new principal and allowed him the autonomy necessary to move forward with his vision. He approached the faculty and said, "this is what is going to work, and this is what we are going to do." Although his approach may be viewed as coming from the top down, this administrator sought teacher buy-in by involving the faculty in the planning and implementation of the grant. A geography teacher in the Ninth Grade Community noted that "he really got this up and running. He wrote a grant and received it. He had done some extensive research into the smaller learning communities...and educated us, and facilitated in a very effective manner, I thought, in which we could be very involved in piecing it together."

A teacher in the Business, Arts, Computers, and Humanities Community recognized that the principal "let us make the choices. He let us have the autonomy of how we were going to divide it (the school) up and stuff like that." When a mathematics teacher in the Ninth Grade Community was asked how he felt about the administration's approach, he said, "I feel like the administration has done a really good job of directing us in the way we need to go."

This sense of allowing teachers to construct and inform the direction of the school was met with great anticipation among some teachers, indifference by other teachers, and disdain by still others. All of which suggests varying levels of commitment regarding teacher buy-in. Some teachers felt that they should be told what they were supposed to do and lamented that they were not told exactly how to accomplish what the principal had envisioned.

At the outset there was some division among the faculty regarding the school's reorganization into smaller learning communities. There were those like an English teacher in the Applied Science and Technology Community who supported the venture. "I bought into it right away, and most of the guys that I work with, even the ones that aren't here anymore." As this teacher saw it, it was the chairperson of the community that obtained this level of support. He was an individual that "we wanted to be in the position. He's done a good job for us. You know, we actually got together. We worked as a team. There were no ego problems or any of that other kind of stuff that interferes with things like this. And so we just had everything ready to go." Another English teacher, who chaired the Ninth Grade Community, was also a supporter of smaller learning communities. "The first year, I was really excited from the get go. It sounded wonderful to me, and I jumped in with both feet."

There were others, however, who were less enthusiastic. Some of the teachers to whom we talked saw this as outright resistance. Most, however, claimed that this was not so much opposition but hesitancy. As one English teacher in the Business, Arts, Computers, and Humanities Community noted there were people in the school "dragging their feet." As this teacher saw it, "there is no faculty that jumps on board to anything if they have people that have been around long enough because they have already seen a lot things and they see them come and they see them go, and so I don't think anybody who has been around a long time gets very eager at first to make changes." Another English teacher in the same community also noted a tendency on the part of the faculty for "dragging their feet" in introducing smaller learning communities. As she saw it, however, those "who dug their heels in" initially began as they saw the benefits of this reform to change their minds."

For a number of teachers at Timberton, smaller learning communities were just one more reform effort that they had undertaken over the years. Such initiatives, according to these teachers, come and go and over time have engendered the attitude in many of them, according to a history teacher in the Health Science and Human Services Community, that "this too shall pass." What bothered these teachers was their belief that the introduction of smaller learning communities would change what had become long standing and comfortable practices at the school. As the chair of the community noted, the response of many teachers was not so much resistance as resignation to its inevitability but also to its ultimate failure.

"We've been through—we've had a couple of different grants. And you know, we don't want to start—do something with one grant, and then that stops, so we are going to do [something] totally different because this was the big thing when we started this whole thing when [the new principal] came here. Many of the teachers that had been here for years, you know, we have done this before—every time we get a new principal—it is a new thing, and you know we are always changing and... this is just one more thing. And you know we will just do it because we have to, but we know it will go away."

According to the drama teacher in the Business, Arts, Computers, and Humanities Community, there were those among the faculty who did not like this initiative and did not see why the school was embarking on it. He went on to say, however, that they recognized that this reform was proceeding ahead with or without them, and they reluctantly accepted it. For some the division was between the school's veteran teachers and those who were newcomers to the faculty. The chair of the Business, Arts, Computers, and Humanities Community put it this way. "I think that there are some teachers that don't want to change, that have taught the same way for twenty years, and you know it is too difficult to try cooperative learning or to try, you know, differentiated instruction, or project-based learning. But I think especially with the newcomers coming in, we are trying to stress that, so hopefully as the old goes out and the new comes in, we have got better ways to teach kids without the traditional methods."

Despite these doubts on the part of some teachers, there seemed to be sufficient support for the reorganization. The assistant principal noted that he didn't see "any negativity." "I see some passive feelings about it a little bit, but most of them are pretty supportive." As the principal saw it, there was about "85 to 95 percent buy-in" on the part of the faculty. In this vein, the chair of the Business, Arts, Computers, and Humanities Community noted that Timberton's teachers were on balance a very cooperative faculty. "Our personality," she noted, "is okay tell us what we are going to do and we will do it." She went on to say that the personality of the faculty at the city's other high school was "resistance" to the reform. There was in fact a deep sense of commitment among Timberton's faculty to the school. A teacher who had been at Timberton Central for seventeen years stated, "I love [Timberton] because of the community. It has a great diversity of kids. I would never want to teach at another high school; I love it here." This sense of commitment was widely felt among faculty and staff as they articulated the smaller learning communities mission, vision, and objectives.

VI.

Timberton Central's purpose for establishing its smaller learning communities initiative was to (1) "improve overall student achievement, thereby decreasing the achievement gap;" and (2) "provide a more personal, productive, and safe learning environment where all students can realize their potential." The goals that assisted in accomplishing this mission were to:

1) improve every student's academic performance to meet high standards of excellence;
2) prepare each student to successfully transition to the next stage* of his/her development (*Each 9th grader for high school; each 10th–12th grader for higher education or workforce);
3) improve student school-related attitudes and behaviors resulting in a more personal, productive, and safe learning environment; and
4) build [Timberton High's] capacity to create, support, and sustain more personal learning communities for students.

The intended outcome of the Timberton's smaller learning communities reform included the desire to "improve the knowledge, skills, attitudes, and behaviors of students, their families, teachers, counselors, and the community."[42]

An English teacher in the Applied Science and Technology Community stated that in his mind the primary objective of the smaller learning communities initiative was to "be able to offer the kids instruction, modified instruction in the core curriculum, that actually is interest-based. And it's not just to keep them here. But it is to prepare them for post-secondary education, even those kids that might not think that they have a chance for post-secondary education." Another faculty member, a mathematics teacher in the Ninth Grade Community echoed these same sentiments by saying that "the goals of the project are to prepare students to go—first of all to graduate from high school with a focus for what they want to do later in life. And then not only to have them graduate, but to prepare them to go on to a higher institution of learning—whether it be a technical school, or the university or college."

Another objective that was not explicitly stated in the formal mission of the grant but was expressed by numerous faculty members was to address the school's changing demographics. Timberton had experienced an influx of Latino/a students who came from low socio-economic households. This has presented a challenge to Timberton in the form of an increasing dropout rate and a growing achievement gap between White and Latino/a students. The Timberton faculty felt that the smaller learning communities initiative could help address those academic ills. The chair of the Health Science and Human Services Community pointed out in that regard that "we have a growing population of Hispanics, and a growing population of ESL students. One of the things we thought smaller learning communities could do specifically for those types of demographics is give them a connection with a small group of teachers—because our biggest problem is getting them to connect with the school, and wanting to come to school. I'm sure without knowing the numbers, but that would be our highest area of dropouts, I'm sure. And we have a hard time getting their parents to be involved in the school, and so a lot of the small school research shows that if you can keep the groups of kids together and give them an interest, they are more likely to continue to come to school."

A history teacher in the same community was, however, less optimistic about the impact of this initiative on the school's Latino/a students. "Well, I think my concern right now is that the small learning community somehow is not stimulating our, I guess you would call them an ESL type of student, your Hispanic kids. I wish we could find a way in the smaller learning community to stimulate their interest to a point that they want to succeed. I don't teach AP history. I teach U.S. History to kids that don't want to take AP history, and so I have a high number of Hispanics in there and my goal is to try to somehow stimulate those kids to want to be there and to do something. Too many of them don't want to do anything. And I still don't see the smaller learning community picking up on that kind of kid. And I don't know what it is going to take to do that, but I don't feel the success in education that I used to because there are so many kids and they are usually Hispanic kids that they are not stimulated somehow to learn. Or they are not involved in the learning process, and I think we are still missing them."

A parent of a Timberton student who worked for the city as an economic planner noted a host of problems including the growth of youth gangs, White-Latino/a racial conflict, and the absence of good employment opportunities that was responsible for the city's low level of educational attainment and high dropout rates among its high school students.

This idea of connectedness and building relationships has become the hallmark of many smaller learning communities reform endeavors. Timberton Central desired through their reform efforts that relationships would be improved and strengthened among students with other students, students with teachers, teachers with students, and teachers with teachers. An English teacher in the Business, Arts, Computers, and Humanities Community noted that "the main thing about small learning communities is that they help foster a relationship between teachers and their students, instead of having a bunch of kids... that you don't really know that well, breaking it down into a smaller group so that you can really get to know them, and then you know how to teach them. It is easier to teach kids you know."

One of the basic premises behind smaller learning communities is that students will have the same teacher for multiple semesters and perhaps multiple academic years. The chair of the Business, Arts, Computers, and Humanities Community expressed her delight in having the opportunity to see the same students by reflecting upon one particular student, "I had a student who was taking a second year class. I had him two trimesters the first year, and I had him two trimesters the second year, and he joined our Future Business Leaders of America, the FBLA program, and he did so well, he went all the way to nationals... and you know he told me at the end of the year, 'I never would have joined this had it not been for you.' And he said, 'I never would have gone to nationals, I would never know this much about computers.' They have the opportunity to be in class and to learn as much as they want. And so when you have a kid over and over like that, and they feel comfortable in your class, and with who you are, it is just a good environment for learning. They can learn as much as they want, and that is what this kid did. He was just given a nice, safe environment, where he knew he could learn as much as he wanted, and he just really excelled, and I just don't think you'd get that in a larger high school, or if they are taking a whole bunch of different teachers. I think kids connect with one or two teachers."

When asked about the purpose of smaller learning communities, a parent to whom we talked echoed the same sentiment. The point of smaller learning communities, as he saw it, was "to foster some sort of community feeling so that there is some synergy [and] learning would occur." Teachers have also been able to collaborate more freely within their community, and as one of them suggested, "the kids know that the faculty get along great. They know that we talk; they know that we support each other; and I think they feel that. And I think that is almost as important to learning as the student's feeling of community, is the sense of faculty that we are all on the same page and that we are working together as a team. That's Timberton's... strongest point, is we have a great faculty and we have a great faculty that gets along well."

Timberton Central has experienced some of the benefits of closer teacher and student interaction. A number of faculty members mentioned the increased collaboration among teachers as a result of smaller learning communities. One teacher who taught mathematics in the Ninth Grade Community mentioned the familial relationships that were forming. "You are in a family, so you discuss any problems, any insights that we notice about a particular student or students. We bounce ideas back and forth, and try to come up with various types of plans to meet the needs of those who are falling behind."

The faculty was given opportunities within their smaller learning communities to have common planning periods and preparation times so that they could collaborate in working with students. The principal also facilitated community meetings to discuss students who were failing at midterm with the intent of articulating a plan to help each student pass by the end of the term. In one instance the principal singled out a teacher who at midterm had 87 percent of his students failing to urge him to change his teaching practices to bring them more into line with the purposes of smaller learning communities. Students have also become more comfortable with their teachers. As one teacher's comments made clear, student-teacher relationships improved in the smaller learning communities. "You have them (students) all year, and it gives you plenty of time to build a good rapport with those kids, a good relationship. I don't eat lunch in the faculty room near as much as what I used to, because I have students who come during lunch time, just to chit-chat with me, because we have built a good rapport. And a lot of times they will come and start just talking and then they become comfortable, and they will ask for tutoring help, where maybe if it was done the old way that would not have happened."

Several teachers noticed that the line which is often drawn between educator and student in typical comprehensive high schools became blurred in Timberton's smaller learning communities. A geography teacher in the Ninth Grade Community noted that "you have to be able to interact, not just as teacher to student, but maybe as human to human . . . I don't think they have a lot of humanness in their homes. Some of my kid's parents are in jail." In the same vein, an English teacher in this community commented that "some of these kids are lacking a mother figure, or whatever, and if they can build a connection to [Timberton Central] then they are going to keep going. They are going to succeed."

Students also noted that the smaller learning communities improved relationships. As one student pointed out, those within a community took the same classes together and over time they got to know each other and were more apt to help each other. Similarly they noted that if a teacher got to know a student better, "they are more willing to help you." Not everyone, however, felt this way. One eleventh grade student who was very critical of the long term relationships that smaller learning communities were supposed to foster commented that "it isn't anyone's right to tell you who to bond with." She went on to criticize the practice of looping within smaller learning communities that placed students with the same teacher for more than one year. Such assignments could, she felt, place a student for an extended period with a teacher who had a different learning style.

VII.

Establishing smaller learning communities at Timberton did, as it turned out, pose serious challenges for the school and its staff. One obstacle to reforming Timberton Central was the idea of change itself. Faculty were not comfortable with having to alter what they had been doing for years. As a mathematics teacher in the Applied Science and Technology Community put it, "one of the biggest obstacles, is just changing. I mean, you have taught for years... I've taught for nineteen years in junior high and I've been here for eight years. We've used certain methods and ways to teach, and handouts... So my cumulative work from other years is all centered on teaching generally, and I have to modify both what kind of assignments and what methods to teach, which is like a first year teacher, you have to learn everything over again. And you have to collaborate with other teachers,... find out what they are teaching in the science classes, or what they are teaching in the business classes..."

Another difficulty that interfered with Timberton Central's reorganization was the belief among some teachers that the reform was just not working in the way that it should. It was, for example, a central belief of the smaller learning communities initiative that an important factor in the ultimate success of students was their selection of a community that matched their immediate interests and possible career aspirations. Yet, it was never all that clear as to whether students were very thoughtful in making this seemingly important choice. From our interviews with students, it seemed that many of them selected their communities on the basis of what their friends did rather than on their interests or long term career goals. And in some cases, it turned out that students were assigned to communities for scheduling purposes and not because of any choice on their part. One teacher commented that "I do think that the thing that would improve it the most, is if the kids would just see the importance of it. I don't know if they really understand what we are trying to do..."

For the faculty, the greatest stumbling block to Timberton Central's reorganization was the imminent retirement of the principal who orchestrated the smaller learning communities initiative. As a history teacher in the Health Science and Human Services Community put it, "when he leaves, I think it is going to change. And when we run out of grant money, I think it is going to change. I don't know if we are going to go back to the way things were, or if we are going to try to nurture and branch out from the small learning community... when you lose the head of the organization, I don't know what is going to happen." This perception created some anxiety for faculty members and in some instances affected the extent of their support. As suggested earlier, some teachers felt that this was just another reform effort that would pass with time and was not worth the required energy to adopt.

From the focus groups that we held with students, it seemed that most students saw Timberton's reorganization as a means of helping them chart their future. One eleventh grader noted that its purpose was "to get you started on the kind of job you want." Another student remarked that "it is like a pathway to what classes you want to take." And still another student noted that "it even helps you decide what you want to do if you don't really know what you want to be; you can go into a learning community, and it gives you choices." Yet, students did report obstacles to the implementation of the reform. One student noted that "I didn't even know that I was in" a community. And another commented that although he signed up for a community, he was never given a reason why he should do so.

Even if students understood the purpose of the smaller learning communities and knew where they wanted to be placed, it did not always, at least as they saw it, work out that way. One student noted that the reform was a good idea that would have worked for him if he did not "get stuck in the wrong learning community last year." He reported that although he was interested in computers, he was originally placed in the Health Science and Human Service Community. But he went on to say that once he talked to his counselor, it was easy to change to the correct community. Another student, however, reported that he wanted to change to a different community and followed the counselor's direction to put in a request but that he never heard back about his request. As the students to whom we talked saw it, smaller learning communities were something new, and they were the first group to experience them. As one interviewee put it, there "might have been bumps" along the way but the transition has occurred relatively smoothly.

One of the parents to whom we talked also noted problems that his daughter faced in selecting a smaller learning community. She played in a band and wanted to be in a community with a group of students who were interested in music. The problem turned out to be scheduling and there was little that he or his wife could do to remedy the situation. "We offered as parents, you know to go to the school on her behalf, and in fact, my wife, I believe, did talk to somebody at the school about it, and came away a little dissatisfied with the answer but it is what it is."

As it turned, out students were not completely satisfied with this reform and did voice dissatisfaction with aspects of the initiative. One student noted that assigning students with perhaps related but different career aspirations in the same community, especially if that placed those heading for college with those in terminal programs was problematic. "I kind of think that the learning communities, especially, you know like, no offense, but like the Health Science and Human Services [Community], you know or whatever, it is a lot like hair cutting and stuff. That's good for people, you know, who don't want to graduate from high school and go on to college and stuff, but you know you should have the people that don't want to graduate from college, have them be in a community."

Notwithstanding the difficulties that accompanied the smaller learning communities reform, the school's faculty identified a number of positive accomplishments. The reform according to one of Timberton Central's special education teachers was creating a safer learning environment. "It has helped us to know who the kids are. We've got some gang problems, the last two days, a shooting, a couple of days ago, a stabbing yesterday, and they have been kids from our school, and kids who have either graduated or are suspended from our school. So they are [Timberton] High kids, and so we have had girls walking in, going to hunt down other people. Yesterday the teachers upstairs recognized that they didn't belong... and they picked them out, just fast. Where I don't think a few years ago, we could have done that. I mean, they knew they did not belong upstairs in that community, and they were after them. So there's something there, I think the kids are safer... It is not quite as easy to sneak into the school as it used to be."

There were also, according to some teachers, indicators that student attendance and achievement were improving as a result of the smaller learning communities. A history teacher in the Ninth Grade Community suggested that "our kids' performance is improving. I mean we had to bring these kids up from way, way down. I mean we have got them coming to school now... we have improved that... You can't teach unless you get them here..." He continued by suggesting that the smaller learning communities movement is bringing students back by providing, "them with something that is going to make them want to be here. And that is my perception—my perception is that is how smaller learning communities work. There are a lot of parents who are bringing their kids back from charter schools... I have spoken to kids whose parents have brought them back from charter schools because of this comprehensive school reform."

Another faculty member stated that "graduation rates are up. We have more kids going to school. We have more kids graduating from high school, I think better prepared to go to some type of job." This has resulted in considerable enthusiasm for the reorganization among many faculty members although they realize that "we still have a lot of work to do too." The chair of the Business, Arts, Computers, and Humanities Community pointed out that establishing smaller learning communities "has been good for us." She went on to say that "I think it is good for teachers, you know, teaching us to do things differently than they have in the past. Otherwise, we would all just be in the same rut that we were 17 years ago. And you know the research says a lot of times teachers teach the way they were taught. So you got to be taught something differently. The professional development opportunities have been great. I mean, we have had money to go to conferences and to learn new things. And I think that is really important, and to bring people into us, and to have full days of workshops and implementing them. So I think it has been a good thing. I think as long as the teachers and the administrators keep it going, it will continue to grow and be good."

It is difficult to know what to make of these favorable faculty impressions when we considered them side by side with the reports of Adequate Yearly Progress (AYP) that the Federal government employs to assess the success of this initiative. The school did not achieve AYP in 2005–2006, the first year that its smaller learning communities were in operation. While the percentage of the school's White students performing at the proficient level in language arts and mathematics was adequate to achieve AYP, the percentage of the school's Latinos/as, economically disadvantaged, and limited English proficient students was not. The following year, the school did attain AYP with Whites, Latinos/as, economically disadvantaged, and limited English proficient students all achieving proficient performance in language arts and mathematics. During the next year, however, only White students at Timberton attained AYP with Latinos/as, economically disadvantaged, and limited English proficient students failing to do so.[43]

VIII.

Related to the accomplishments of smaller learning communities was the issue of the impact of this reform on pedagogical and curricular practices. Despite the structural changes that have accompanied the establishment of smaller learning communities, its purpose was not simply to restructure large high schools into smaller units. School size was in fact an intermediate goal that was designed to create the conditions that would allow for changing teaching and learning in ways that would enhance student academic achievement.[44] As the chair of Timberton's Health Science and Human Services Community noted the purpose of the school's reorganization was to increase student achievement. "A small school," she pointed out, "is not the destination; it is the vehicle to get you there."

Faculty were, it turns out, divided about the extent of instructional changes that had taken place with the introduction of smaller learning communities. Some teachers claimed that the initiative had no impact on their teaching or on the content of their classes. A veteran science teacher in the Ninth Grade Community noted that despite the reorganization of the school, little had changed in his classroom. Others, however, saw a range of changes. A history teacher in the Health Science and Human Services Community noted both continuity and change. Although he continued to present his material in the same way as he always did, he pointed out that the establishment of smaller learning communities had brought with it a new emphasis on student writing. And still others pointed to significant changes. A math teacher in the Ninth Grade Community commented that with the introduction of smaller learning communities she did less in the way of lecturing and "more activity based, more projects, learning based techniques."

An English teacher in the Health Science and Human Services Community noted smaller learning communities had increased teacher collaboration. "You know when we first started we were teamed up with a teacher and we still are. And so I would be teamed up with a history teacher because I teach English. And since we have the same kids, we teach, you know, from Egypt, the middle ages—the same time he is teaching that, I am teaching that in English. And so that is a nice lineup. You are not on an island any more. You have a group of other teachers that you work with."

And collaboration brought with it curriculum changes. A science teacher in the Health Science and Human Services Community pointed out how the school's reorganization had altered the English curriculum. Prior to the advent of smaller learning communities he noted that English teachers simply taught English. The English teachers in his community, however, worked cooperatively with health teachers and had to provide instruction in a specialized English that he referred to as "medical English." Similarly, his community provided that student be instructed in "medical math" and "medical history." He went on to say that he would "assign a paper that the student can take to three different classes—their English class, maybe their math class... a history class."

Smaller learning communities, according to an English teacher in the Health Science and Human Services Community created a venue for instructional innovations. She introduced what she called a "Passion

Project" that allowed students to devote a semester to working on a project for which they had a "passion" and which would be a "stretch" for them. At the beginning of the year students would spend the month of September identifying a project and developing a schedule for completing it. During the course of the semester students would complete "milestones" in which they would inform her of the progress that they are making. At the end of the semester in November, students would make a short presentation on their projects and exhibit some visible evidence that they have completed them. The only limitation on these projects, according to this teacher, was the student's imagination. They could range from a study of various foods and how to cook and prepare them to their post–high school plans. For the student, she noted, "it gives you the opportunity to design a curriculum for yourself."[45]

Over the course of the year that we were in the school efforts at curriculum and pedagogical change ebbed and flowed. Initially, there was great interest in the introduction of a junior project as a means of integrating the curriculum around issues of interest to students. The most progress in this direction occurred in the Applied Science and Technology Community where faculty developed a year long plan with accompanying written guidelines that directed students to identify topics of interest to them, to undertake exploratory research, to select a faculty member to advise them, and by the end of the year to write a junior thesis.[46]

Other communities, it seemed, made little headway. We attended meetings of the Health Science and Human Services and the Business, Arts, Computers, and Humanities Communities that occurred during the second year in which they considered the idea of the junior project. Faculty comments and questions at both meetings were centered on the definition of the project, whether it was mandatory or voluntary, its relationship to graduation requirements, and the role of faculty in directing and evaluating it. Their remarks suggested that they had made little progress in the previous year in settling on nature of the junior project and were only now beginning to attend to it. During much of the remainder of the year, interest as well as concrete work on the junior project subsided as the faculty and staff confronted both the approaching retirement of the principal and a district wide reorganization that would move the school's Ninth Grade Community to the city's middle schools that were themselves being transformed into junior high schools.

Toward the end of our stay at Timberton, the faculty's work with consultants from Stanford University's School Redesign Network that had begun earlier in the year seemed to bring about a renewed interest in curricular and pedagogical change. A major initiative in this regard was the appointment of instructional coaches. Originally a volunteer position and later an appointed one, these individuals were teachers at Timberton who were assigned for one class period a day to work with other teachers in the school in improving their instructional practices. The coaches who we interviewed noted that their principal assignment, during the initial year of the program was to assist teachers in working collaboratively and in implementing a program of differentiated instruction. The coaches encouraged teachers to provide alternative content, materials, teaching strategies, and technology to meet the diverse needs of these students. A history teacher whose class we visited in the Business, Arts, Computers, and Humanities Community used differentiation in a study of change between the 1920s and the present. She divided her class into three groups that employed different ways of reporting on cultural changes across the decades. The analytical group developed a timeline of key events and people. A practical group was assigned the task of constructing a poster to depict each of the decades between 1930 and 1980. And a creative group was assigned to identify a popular radio or television show during this period and to reenact it or create a video.

Similarly, a French teacher who we interviewed developed a "Dream Trip to France" project that divided her class into small groups involved in preparing travel brochures for trips to France that emphasized particular features of the country—history, the arts, sports, and foods. Within each group students were assigned different roles including tour guide, tour director, arts and entertainment reporter, copy editor, and advertiser that necessitated different levels of skill and ability and led to different written products.

The changes that we saw during the latter part of our time at Timberton were not changes that could only have occurred in smaller learning communities. The differentiation that we observed could have just as easily taken place in a large comprehensive high school. One of the instructional coaches we interviewed made the point in this regard that it was not the establishment of smaller instructional units at Timberton that was so important. What was crucial, she went on to say, was that the reform had created "a climate of change" at the school that had made all sorts of reforms possible. She noted that in the past teachers did not discuss instructional strategies with each other. She herself commented that she would never have gone to another teacher to inquire how that individual taught. As another instructional coach saw it, the introduction of smaller learning communities had created an environment where teachers talked to other teachers more frequently and were more aware of what they are doing instructionally. She felt that such conversations offered teachers feedback that enabled them to teach more effectively.

Not every teacher at Timberton had this positive view of smaller learning communities. Two teachers who we interviewed expressed strong support for the program when we first met them at the beginning of our stay at the school. By midyear, however, they began to express doubts about the efficacy of the program. One, a geography teacher in the Ninth Grade Community, thought that the program had stalled because of the fact that the principal was leaving the school, lack of communication between teachers, and increasing class sizes. She felt that any collaboration or curriculum integration that had occurred could have taken place regardless of the initiative.

The other individual, an English teacher in the Ninth Grade Community, noted that his classroom was unaffected by the initiative. He identified a number of difficulties that had in his words made the reorganization "non-existent." They included the impending retirement of the principal, a shift of interest away from smaller learning communities to other reforms including differentiation and problem-based learning, little teacher "buy-in," and the feeling that the grant money that supported the effort would soon run out.

Students, it seems, also voiced criticisms. Their overall impression was that the reform had brought with it little in the way of changes and that as a consequence smaller learning communities had no impact on their lives. One student thought that the reorganization would mean that her classes would pertain to her community but as she put it, "our classes are just the normal classes and you don't think—oh these are in my learning community." Another student said that he was told that his mathematics and English classes in the Health Science and Human Services Community would be different and related to the medical or health fields, but they were not. He added that maybe "they just haven't gotten around to it." Similarly, an interviewee pointed out that "it hasn't affected any of my classes that I have taken whatsoever. With or without learning communities, I'd still be in the same classes I am now."

Overall, students thought that smaller learning communities possessed the potential to improve the school. Yet, there were the organizational problems regarding the assignment of students to communities that we have already mentioned. And there was, some students mentioned, the need for teachers themselves to become motivated about the reform and in turn to motivate their students. Doing so, they went on to say, was not easy. Teachers were preoccupied just trying to work with their classes as they were presently constituted. As one student noted, "they are trying to come together and get students to come to class to be able to learn, and you know just adding that extra responsibility of a learning community to that is just an extra burden that they have to focus on."

IX.

As we end this chapter we need to ask ourselves how the concept of community helps us in understanding Timberton Central's smaller learning communities initiative. Like any other conceptual lens that we might use, this one directs us to certain issues over and above others. A particular problem with the notion of community in this regard is that it has been and remains something of a vague and imprecise notion with multiple meanings and real questions as to whether it refers to a real state of affairs. Yet, the concept does point to ideals, beliefs, and visions that are vital and important in our social life.

The resolution to this dilemma that was pursued in this book was to conceptualize the notion as an empty signifier—that is as a word without a single meaning—but comprising multiple discourses that in different ways point to the connections between and among the stakeholders of reform. Such connections require that these stakeholders possess a similar understanding of the reform, embrace the reform as constituting the common good to which their work should be directed, and posses the kind of close and trusting relationships that allow them to work together to achieve its goals. Taken together, these are the attributes of a sense of collective belonging.

At the end of the last chapter, we noted that school reform is a complex process that requires among other things a common understanding among stakeholders about their vision for reform and the goals to achieve it, support for those entrusted with the leadership of reform, and a shared recognition about the obstacles that stand in the way of change. The most striking conclusion that emerges in our look at Timberton's smaller learning communities initiative from the lens of community is how difficult it is within an actual school setting to create the sense of belonging that would allow such reforms processes to be realized. Our interviews at Timberton do provide indicators of a movement among the stakeholders in the direction of a sense of collective belonging. Early on in this effort, a group of Timberton's teachers visited a number of small schools and smaller learning communities around the country. Yet despite their different experiences, they were ultimately able to come together around a common vision for the school and the goals required to realize it. The support of teachers for the initiative as well as the mutual sense of trust that existed between teachers and the school's principal also indicates that the kind of common understanding about this effort was beginning to develop. Also important in this regard was the development of closer relationships between teachers and students, the increased collaboration that was occurring among the faculty, and the beliefs voiced by faculty regarding the improvements that the initiative had brought, particularly to the school's pedagogical and curricular practices. Taken together our interviews suggest that among the stakeholders involved in this reform—administrators, teachers, students, and parents—some level of common understanding and agreement about this initiative had developed.

Yet, having said all of this, our interviews also reveal a clear level of discord that undermines the presence of any sense of collective belonging among the reform's stakeholders. Most important in this respect were the views voiced by a minority of teachers that the smaller learning communities initiative was a misdirected reform that would not outlast the principal's presence at the school, that it had been poorly organized and implemented, and that it had not thus far brought much in the way of change to the school or to their own pedagogical and curriculum practices. Lending credence to these criticisms were the comments of students about their lack of understanding of the purposes of the smaller learning communities initiative, the problems that they encountered in selecting a community, and their sense that these communities had little or no impact on their lives.

There is also other evidence that cast doubt about the extent to which stakeholders developed a shared understanding of this reform. A major impetus for reorganizing Timberton into smaller learning communities was to enhance the academic achievement of low performing students, many of whom were of Latino/a origin. Yet despite this intention, we found it difficult throughout our stay at the school to engage teachers and others in conversations about the interplay between race and the smaller learning communities initiatives. We did cite earlier in this chapter the only two extensive comments that were made about the potential impact of smaller learning communities on the school success of Latino/a students. And those comments were in conflict—one suggesting a positive impact and the other less so.

Another issue related to race is also worthy of comment. We had hoped that in our interviews at the school, we would have a chance to talk to students of color about their experiences with smaller learning communities. Since this reform was introduced in large part to address issues related to their academic achievement, we thought that it would be important to obtain their views. We did hold several focus groups comprised of students from the student council and from two English classes whose teachers were willing to help us in making selections and securing parental permission. The majority of these students were White. In hopes of sampling the opinions of a more diverse array of students, one of us stationed himself around the noon hour outside the main office in an open area where small groups, many of whom were students of color, congregated to visit and talk. Standing in the midst of these groups of students, he asked whoever would talk to him about their experiences with smaller learning communities. Their unanimous response to this question was that they had never heard about these communities and had no involvement with them. Clearly, this was not anything approaching an objective sampling. It does, however, raise questions about whether a common understanding of this reform had developed among a wider group of students including those who were non-White.

The disagreement about the impact of smaller learning communities on the instructional program also points to a lack of common understanding about this reform. One of our goals during the year that we spent at Timberton was to observe actual classes in session to ascertain for ourselves the impact of the reform on teaching and learning. It was only at the end of the year that one of us was in fact able to observe a couple of classes. Earlier, teachers were unwilling or at least reluctant to invite us into their classes on the grounds that either the initiative had not affected their practice or that it was too soon to see any changes. We were particularly excited about the prospect of visiting the classes of two teachers in the Ninth Grade Community who had an extensive record of collaboration out of which they had developed a curriculum that integrated science and geography. These teachers were receptive us to our request and suggested that we wait until after the Christmas vacation to visit.

When the school reopened after the break we approached the teachers to find that they had withdrawn their invitation for our visit. As one of them told us, there was no point observing because they were now involved in preparing the students for the end of year state competency tests and had no time to do the "fun things" that they had developed collaboratively in their smaller learning community. Their response perplexed us in that the very point of this reform was to spur forward collaborative efforts from teachers to develop integrated curricula that would enhance student performance on mandated competency tests. We were surprised that these two teachers who were very supportive of Timberton's smaller learning communities initiative would characterize the work that they had done in this new environment as constituting "fun" activities and not relevant to preparing their students for their end of year testing. We wondered if these teachers had the same understanding of this reform as did Timberton's principal and district administrators who introduced and supported it.

We returned to Timberton Central in the spring of 2009 to see what had occurred in the almost two years since we had concluded our study of the school's reorganization. A number of major alterations had taken place. The organizational structure of the communities was solidified with each unit having an administrator, a secretary, and a counselor. Department chairs were given a clear role in coordinating curriculum across communities in the different subject areas. And efforts were underway to transform the faculty of each unit into a professional learning community.

The instructional changes that the staff was beginning to introduce between 2006 and 2007 were now in full swing. Each of the communities had a common planning period to enable teachers work collaboratively. The school had made significant progress in teacher assignments so that most faculty were able to teach all of their courses within a single community. The communities had their first experience in instituting an integrated writing project. And increasing number of students seemed to be aware of the school's reorganization as well as their assignment in one of the school's learning communities.

Clearly Timberton had not made all of the changes that were anticipated when we arrived at the school in 2006. The improvements in academic achievement that had been the impetus for this reform did not seem to have occurred. Yet, using a concept of community as a lens for examining Timberton Central's smaller community initiative tells us that this was not an unsuccessful reform. It had for the most part improved student-student and student-teacher relationships as well as enhanced collaboration among faculty—neither being small achievements in their own right. It was, however, an effort that did not develop a sense of collective belonging among its stakeholder or activate the level of civic capacity that we argued in chapter one as being crucial for successful urban school reform. In that sense, Timberton's smaller learning communities exhibits the mixed record of success and failure that seems, at least from the case studies in this volume, to typify urban school reform since 1960.

A notion of community does tell us something about our recent efforts to reform urban schools. Clearly it does not answer every question we might have, but does answer some, particularly those concerning the relationships and connections among stakeholders. One wonders, however, how useful this notion can be for interpreting new and emerging efforts to reform urban schools. Does a notion of collective belonging, as it has been developed it in this book, provide the kind of conceptual lens that we need in the globalized world that we are fast becoming? Specifically has the changing role of the state that we previewed in chapters five and six from one of direct governing to that of enabling call for a different conceptual framework for understanding school reform? These are issues that will be considered in the epilogue to this volume.

Epilogue

Community in a Cosmopolitan World

In the introductory chapter of his recent book, *The Outliers: The Story of Success*, the journalist Malcolm Gladwell relays the account of one Dr. Stewart Wolf, a physician and professor of medicine at the University of Oklahoma, regarding the remarkable health of the inhabitants of the Borough of Roseto, Pennsylvania. According to Wolf, who often spent his summers during the 1950s on a farm in the vicinity of the borough, Roseto's population was virtually free of individuals under the age of sixty-five who had heart disease. The absence at the time of cholesterol lowering drugs and medical techniques to prevent heart attacks made this an unusual situation. His examination of the residents' medical records along with the results of blood tests and EKGs eliminated most of the common medical explanations for the borough residents' good health, namely genetic makeup, diet, and exercise. The principal cause of death in the borough turned out to be old age.[1]

As he and a colleague conducted their study of the residents, Wolf noted that over the years the population had established patterns of close personal contacts and strong relationships that created an egalitarian social structure within Roseto and sheltered its residents from an array of external pressures. He noted, for example:

> ...how the Rosetans visited one another, stopping to chat in Italian on the Street, say, or cooking for one another in their backyards. They learned about the extended family clans that underlay the town's social structure. They saw how many homes had three generations living under one roof, and how much respect grandparents commanded. They went to mass at Our Lady of Mount Carmel and saw the unifying and calming effect of the church. They counted twenty-two separate civic organizations in a town of

> just under two thousand people. They picked up on a particular egalitarian ethos of the community, which discouraged the wealthy from flaunting their success and helped the unsuccessful obscure their failures.[2]

It was, he argued, their unity, solidarity, and a sense of commitment to the common good that made them healthy.

For Gladwell, the case of Roseto represented an important change in our understanding of personal health and well being. It shifted our attention away from medical and biological explanations to one involving community. To know why the residents of this town were healthy, required Wolf and those who assisted him to, in Gladwell's words, "look beyond the individual. They had to understand the culture he or she was part of, and who their friends and families were and what town their families came from. They had to appreciate the idea that the values of the world we inhabit and the people we surround ourselves with have a profound effect on who we are."[3]

Gladwell's chapter was in effect pointing to the important role that a notion of community plays in helping us interprets cultural practices, in his case community health and in ours urban school reform. The question that we need to pose in this epilogue is the continuing value of the notion of community for such purposes. My concern in this regard lies with the fact that the concept of community is typically used to refer to those discourses that circulate in the conceptual space surrounding the idea of nationhood. These are discourses that act to convey a sense of collective belonging to any of a number of entities. One such entity is the state, which as I write is undergoing significant transformations.

These changes were discussed in chapter six through the lens offered by proponents of "third way" thinking. One element of this shift involves the accelerating pace of globalization, which I defined in an earlier chapter as the free movement of capital, labor, and communication across national borders. Under these conditions, a single nation can no longer be the sole player in the major economic and political issues affecting it. External, international forces are involved, either other nations or transnational government or nongovernmental agencies of one sort or another.

A second element of this shift is a change in the role of the state. For proponents of "third way" thinking, this has meant a shift from that of directing to one of enabling.[4] While the state ceases to govern directly, it does not retreat from its role in regulation. Rather, that role occurs indirectly and is mediated by civil society operating through such nongovernmental organizations as public private partnerships. At the same time, the mechanism for regulating individual conduct shifts from external social control to self-governance. According to the sociologist Nikolas Rose:

> The state is no longer to be required to answer all society's needs for order, security, health and productivity. Individuals, firms, organizations, localities, schools, parents, hospitals, housing estates must take on themselves—as 'partners'—a portion of the responsibility for resolving these issues—whether this be by permanent retraining for the work, or neighbourhood watch for the community. This involves a double movement of autonomization and responsibilization. Organizations and other actors that were once enmeshed in the complex and bureaucratic lines of force of the social state are to be set free to find their own destiny. Yet, at the same time, they are to be made responsible for that destiny, and for the destiny of society as a whole in new ways. Politics is to be returned to society itself, but no longer in a social form: in the form of individual morality, organizational responsibility and ethical community.[5]

How that reconstruction plays itself out is, however, less certain. As I write, the current worldwide economic recession has resulted in a number of state driven and funded initiatives to rescue both nations and individuals from economic collapse. Such policies seem to point to something of a reversal of this move from a more indirect state role toward one that is directive and intrusive. At the same time, however, the push toward privatization and choice, the hallmarks of neoliberalism, remain strong.[6] In this environment, community, if it is to be a useful conceptual framework, requires something of a recalibration. And doing so brings us to the concept of cosmopolitanism.

II.

In our initial discussion of the notion of community in chapter one, we pointed to its role as conveying a sense of belonging that binds groups and individuals together around common goals and a sense of the common good. In subsequent chapters, we used this notion of community as a lens for understanding those cultural practices associated with urban school reform. For secondary education, particularly, we have seen how looking at the high school through the framework of community can enhance our understanding of how that institution works and the challenges that it faces.[7]

The shift to a globalized world does not negate the need for such a conceptual framework for not only secondary education bur for schooling in general. If anything, the diversity of such a world and the myriad of complex problems that it poses for those of us who inhabit it heighten the need for such a lens. Yet the scope of the lens must be different, and a notion of cosmopolitanism provides for that difference.

Like the notion of community, cosmopolitanism can be seen as a floating signifier with a diversity of competing meanings. Undergirding these multitude meaning and constituting the great virtue of the idea of cosmopolitanism is that it points to a unified and collective sense of belonging that binds people together and cuts across the local settings in which they live and work. It is a notion that stands against such particularities as race, gender, and nationhood.[8] Cosmopolitanism, then, challenges the conventional wisdom that supreme power lies in the hands of the state and entrusts it instead to a notion of "world citizenship."[9] It is a notion, according to the philosopher Martha Nussbaum, that calls on us to offer our principal loyalty "to the moral community made up by the humanity of all human beings."[10]

Yet, under conditions of globalization, states have hardly disappeared Despite the shift from their role of directing to that of enabling, their place vis-à-vis transnational and international institutions ebbs and flows with differing circumstances. What is called for, then, is a versatile discourse through which we can accommodate different demands for solidarity and commonality, some that are particular and operate at the level of the state and others that are universalistic and encompass all of human kind. With its roots in ancient Greek and eighteenth century Enlightenment thought, the idea of cosmopolitanism provides us with the language that we need to shape our notion of community in those twin directions.

One of the dilemmas with the notion of community that we have considered throughout this volume has been its multiple and often uncertain meanings. In chapter one, we identified two very distinct discourses for talking about the idea of community. One, which we identified with a conservative intellectual tradition, defined community in static and exclusionary terms to refer to a host of characteristics that separate individuals and groups from each other. Such a notion of community has been used to limit group membership to certain individuals possessing certain supposed desirable characteristics while at the same time constraining, isolating, and even eliminating those lacking those same characteristics. The other, more liberal notion defined community in inclusionary terms to talk about the similarities among individuals and groups that bind them together. It is a discourse that has been used to promote equality and democracy.

As we search for a notion of community for interpreting the cultural practices of a globalized world, it is clear that what is called for is a more liberal and inclusive notion of this concept. There are certain features of cosmopolitan discourse—its devotion to rationality and science, its belief in humans freedom, and its rejection of dogmatism and mysticism—that can construct and shape our institutions as well as ourselves in desirable ways.[11] There is, however, the danger that cosmopolitanism like community can point us in the wrong direction. To the extent that this notion serves to include certain individuals, namely those who exhibit a scientific and rational outlook, it allows for the exclusion of those who do not possess these characteristics. Ultimately, then, a notion of cosmopolitan community can provide us with an appropriate conceptual lens for interpreting school reform in a globalized world, or it can also lead us astray. But that is the subject for another book.

Notes

1 Community and Curriculum: A Conceptual Framework for Interpreting Urban School Reform

1. Timberton is a pseudonym for the real community and school in which the research described here and in chapter 7 was conducted. Those participating in the interviews on which this research was based were guaranteed anonymity and are therefore not identified.
2. "Improving Student Achievement through Smaller Learning Communities" Grant Application, 2003; SLC and Core Discipline Agenda, January 24, 2007.
3. Robert MacIver, *Community: A Sociological Study* (London: Macmillan, 1917), 81, cited in Barry M. Franklin, *Building the American Community: The School Curriculum and the Search for Social Control* (London: Falmer Press, 1986), 8.
4. See, for example, Thomas S. Popkewitz, *Cosmopolitanism and the Age of School Reform: Science, Education, and Making Society by Making the Child* (New York: Routledge, 2007).
5. Barry M. Franklin, *From "Backwardness" to "At-Risk": Childhood Learning Difficulties and the Contradictions of School Reform* (Albany: State University of New York Press, 1994); Barry M. Franklin, "Creating a Discourse for Restructuring in Detroit: Achievement, Race, and the Northern High School Walkout," in *Educational Restructuring: International Perspectives on Traveling Policies*, ed. Sverker Lindblad and Thomas S. Popkewitz (Greenwich: Information Age, 2004), 191–217; Barry M. Franklin, "State Theory and Urban School Reform I: The View from Detroit," in *Defending Public Education: Schooling and the Rise of the Security State*, ed. David A. Gabbard and E. Wayne Ross (New York: Teachers College Press, 2004), 117–129
6. Barry M. Franklin, Marianne N. Bloch, and Thomas S. Popkewitz, "Educational Partnerships: An Introductory Framework," in *Educational Partnerships and the State: The Paradoxes of Governing Schools, Children, and Families*, ed. Barry M. Franklin, Marianne N. Bloch, and Thomas S. Popkewitz (New York: Palgrave Macmillan, 2003), 1–23.
7. Derek L. Phillips, *Looking Backward: A Critical Appraisal of Communitarian Thought* (Princeton: Princeton University Press, 1993), 3.

8. Susan L. Hodgett, "Community, Sense Of," in *Encyclopedia of Community*, ed. Karen Christensen and David Levinson (Thousand Oaks: Sage Publications, 2003), 2: 236–237.
9. Raymond Williams, *Keywords: A Vocabulary of Culture and Society*, rev. ed. (New York: Oxford University Press, 1983), 75–76.
10. Alan Ehrenhalt "Where Have All the Followers Gone?" in *Community Works: The Revival of Civil Society in America*, ed. E.J. Dionne Jr. (Washington, DC: Brookings Institution Press, 1998), 93.
11. Robert Booth Fowler, *The Dance With Community: The Contemporary Debate in American Thought* (Lawrence: University Press of Kansas, 1991), 39–41.
12. Nikolas Rose, *Powers of Freedom: Reframing Political Thought* (Cambridge: Cambridge University Press, 1999), 172.
13. "McDonald's and You," 2003.
14. "Starbucks in our Communities," 2006.
15. *Zions Bank Community Magazine*, Winter, 2007.
16. Phillips, 7; Lynn Fendler, "Others and the Problem of Community," *Curriculum Inquiry* 36 (Fall, 2006), 303–326.
17. Suzanne Keller, *Community: Pursuing the Dream, Living the Reality* (Princeton: Princeton University Press, 2003), 4–7
18. George A. Hillery, Jr., "Definitions of Community: Areas of Agreement," *Rural Sociology* 20 (June, 1955), 111–123.
19. Benedict Anderson, *Imagined Communities*, rev. ed. (London: Verso, 1991), 4.
20. Ibid., 6.
21. Phillips, 12–14.
22. Ibid., 29–36.
23. Ibid., 149.
24. J.G.A. Pocock, *The Machiavellian Moment: Florentine Political Thought and the Atlantic Republican Tradition* (Princeton: Princeton University Press, 1975), Parts 1 and 2.
25. Michael J. Sandel, *Democracy's Discontent: America in Search of a Public Philosophy* (Cambridge, MA: Harvard University Press, 1996), 5–8, 133–200.
26. For a discussion of this issue see Alasdair MacIntyre, *After Virtue* (Notre Dame, IN: University of Notre Dame Press, 1981).
27. William Davies, "Against Community," *Prospect* (November, 2006), 15–16.
28. Saint Augustine, *City of God*, trans. Henry Bettenson (London: Penguin Books, 2003), 73–75, 881–883.
29. Charles Marsh, *The Beloved Community: How Faith Shapes Social Justice, from the Civil Rights Movement to Today* (New York: Basic Books, 2005), 49–50, 207–210; Kenneth L. Smith and Ira G. Zepp, Jr., *Search for the Beloved Community: The Thinking of Martin Luther King, Jr.* (Valley Forge: Judson Press, 1974), 119–140.
30. Marcus Raskin, *The Common Good: Its Politics, Policies, and Philosophy* (New York: Routledge and Kegan Paul, 1986), 26–27; Amitai Etzioni, *The Common Good* (Cambridge: Polity Press, 2004), 1–2; Anthony S. Bryk, Valerie E. Lee,

and Peter B. Holland, *Catholic Schools and the Common Good* (Cambridge, MA: Harvard University Press, 1993), 38.

31. Cory Booker, interview by Bill Maher, *Real Time*, Home Box Office, November 7, 2008.
32. Kathy Greeley, *Why Fly That Way: Linking Community and Academic Achievement* (New York: Teachers College Press, 2000).
33. Ibid., 113–114.
34. Arnold F. Fege, "Getting Ruby a Quality Public Education: Forty-Two Years of Building the Demand for Quality Public Schools through Parental and Public Involvement," *Harvard Educational Review* 76 (Winter, 2006), 570–586.
35. Ibid., 580.
36. Raymond Plant, *Community and Ideology: An Essay in Applied Social Philosophy* (London: Routledge and Kegan Paul, 1974), 8.
37. Ibid., 9.
38. Frank Fischer, *Reframing Public Policy: Discursive Politics and Deliberative Practices* (Oxford: Oxford University Press, 2003), 73–76; Maarten A. Hajer, *The Politics of Environmental Discourse: Ecological Modernization and the Policy Process* (Oxford: Oxford University Press, 1995), 41–46. In developing my understanding of the notion of discourse, I am indebted to ongoing conversations that I have had with three colleagues at Utah State University—Steve Camicia, Cinthya Saavedra, and Patricio Ortiz.
39. Ernesto Laclau, "Why do Empty Signifiers Matter to Politics?" in *The Lesser Evil and the Greater Good: The Theory and Politics of Social Diversity*, ed. Jeffrey Weeks (London: Oram Press, 1994), 167–178.
40. R. Buenfil Burgos, "Partnership as a Floating and Empty Signifier within Educational Policies: The Mexican Case," in Franklin, Bloch, and Popkewitz, 55.
41. I am indebted to Thomas Popkewitz for this way of viewing the notion of community. See *Cosmopolitanism*, 68.
42. Plant, 25.
43. Phillips, 164–176; Philip Selznick, *The Moral Commonwealth: Social Theory and the Promise of Community.* (Berkeley: University of California Press, 1992), 360.
44. Robert D. Putnam, *Bowling Alone: The Collapse and Revival of American Community* (New York: Simon & Schuster, 2000), 22–24, 357–358. Putnam attributes the origins of the concept of social capital to Lyda Judson Hanifan, an early twentieth century supervisor of rural schools in West Virginia. As Hanifan saw it, social capital referred to cultural practices that enabled individuals to effectively utilize the array of material resources that they had available to them. Such practices included in his words "good will, fellowship, mutual sympathy, and social intercourse." See L.J. Hanifan, "The Rural School Community Center," *Annals of the American Academy of Political and Social Science* 67 (1916), 130.
45. *Building the American Community*, Chapters. 3–4.
46. Putnam, 273–275.

47. Ibid., Section II.
48. Ibid., Section V.
49. Robert D. Putnam and Lewis M. Feldstein with Don Cohen, *Better Together: Restoring the American Community* (New York: Simon & Schuster, 2003).
50. Clarence N. Stone, Jeffrey R. Henig, Bryan D. Jones, and Carol Pierannunzi, *Building Civic Capacity: The Politics of Reforming Urban Schools* (Lawrence: University Press of Kansas, 2001), 4.
51. Clarence N. Stone, *Regime Politics: Governing Atlanta, 1946–1988* (Lawrence: University Press of Kansas, 1999), vii, 9–19.
52. Clarence N. Stone, "Civic Capacity: What, Why, and from Whence," in *The Public Schools*, ed. Susan Fuhrman and Marvin Lazerson (New York: Oxford University Press, 2005), 231; Eva Gold, Elaine Simon, Maia Cucchiara, Cecily Mitchell, and Morgan Riffer, *A Philadelphia Story: Building Civic Capacity for School Reform in a Privatizing System* (Philadelphia: Research for Action, 2007), 2.
53. Stone, Henig, Jones, and Pierannunzi, 4–12; Clarence N. Stone, "Civic Capacity and Urban School Reform," in *Changing Urban Education*, ed. Clarence N. Stone (Lawrence: University Press of Kansas, 1998), 250–273; Clarence N. Stone, "Civic Capacity and Urban Education," Unpublished Paper, retrieved May 15, 2008 from http://www.bsos.umd.edu/gvpt/stone/prolo.html/.
54. I am indebted to Daniel Tröhler for making this distinction between a moral and analytic definition of community and for helping me to see why the latter makes more sense for this kind of study.
55. Jośe R. Rosario, "Communitarianism and the Moral Order of Schools," in *Curriculum and Consequence: Herbert M. Kliebard and the Promise of Schooling*, ed. Barry M. Franklin (New York: Teachers College Press, 2000), 37–42.
56. Lea Hubbard, Hugh Mehan, and Mary Kay Stein, *Reform as Learning: School Reform, Organizational Culture, and Community Politics in San Diego* (New York: Routledge, 2006), 1–4, 41–44.
57. Ibid., 183–237. Larry Cuban and Michael Usdan's earlier study of this reform initiative paints a similar picture of conflict. See Larry Cuban and Michael Usdan, "Fast and Top Down: Systemic Reform and Student Achievement in San Diego City Schools," in *Powerful Reforms with Shallow Roots: Improving America's Urban Schools*, ed. Larry Cuban and Michael Usdan (New York: Teachers College Press, 2003), 77–95.
58. Hubbard, Mehan, and Stein, 11.
59. Ibid., 16–17.
60. Ibid., 87.
61. Ibid., 208–209.
62. *New York Times*, September 24, 2007.
63. Bruce Fuller, "The Public Square, Big or Small? Charter Schools in Political Context," in *Inside Charter Schools: The Paradox of Radical Decentralization*, ed. Bruce Fuller (Cambridge, MA: Harvard University Press, 2000), 20–22.

64. Bruce Fuller, "Breaking Away or Pulling? Making Decentralization Work," Ibid., 236, 247–248.
65. Fuller, "The Public Square," 27.
66. John Dewey, "The School and the Society," in *John Dewey: The Middle Works, 1899–1924*, ed. Jo Ann Boydston (Carbondale: Southern Illinois University Press, 1976), 1: 7.
67. John Dewey, *The Public and Its Problems* (New York: Henry Holt), 110–142.
68. Ibid., 147.
69. Robert B. Westbrook, *John Dewey and American Democracy* (Ithaca: Cornell University Press, 1991), 293–300; Alan Ryan, *John Dewey and the High Tide of American Liberalism* (New York: W.W. Norton, 1995), 216–217. For Lipmann's critique of democracy see Walter Lippmann, *Public Opinion* (New York: Harcourt, Brace, 1922), 3–32, 379–410 and Walter Lippmann, *The Phantom Public* (New York: Harcourt, Brace, 1925), 143–186. For Dewey's critique of Lipmann's views see his review of *Public Opinion* in *John Dewey: The Middle Works*, 13: 337–344 and his review of *The Phantom Public* in *John Dewey: The Later Works, 1925–1933*, ed. Jo Ann Boydston (Carbondale: Southern Illinois University Press, 1988), 2: 213–220
70. Review of *Public Opinion*, 13: 337.
71. John Dewey, "The Ethics of Democracy," in John Dewey: *The Early Works, 1882–1898*, ed. Jo Ann Boydston (Carbondale: Southern University Illinois Press 1969), 1:232–234.
72. Ibid., 237.
73. John Dewey, "The Reflex Arc in Psychology," in *John Dewey: The Early Works*, 5: 99.
74. Ibid., 102.
75. I discuss the challenge that a notion of reciprocity poses to behaviorism in comparing the psychological views of Edward L. Thorndike and George Herbert Mead. See *Building the American Community*, Chapter 3.
76. John Dewey, *Democracy and Education* (New York: Free Press, 1966), 4–5.
77. John Dewey, *Freedom and Culture* (New York: G.P. Putnam's Sons, 1939), 12.
78. Westbrook, 303–304; James Campbell, "Dewey's Conception of Community," in *Reading Dewey: Interpretations for a Postmodern Generation*, ed. Larry A. Hickman (Bloomington: Indiana University, 1998), 32–35.
79. Robert N. Bellah, "Community Properly Understood: A Defense of 'Democratic Communitarianism," in *The Essential Communitarian Reader*, ed. Amitai Etzioni (Lanham: Rowman & Litttlefield, 1998), 16.
80. Veronica Garcia, Wilhemia Agbemakplido, Hanan Abdella, Oscar Jopez, Jr, and Rashida T. Registe, "High School Students' Perspectives on the 2001 No Child Left Behind Act's Definition of a Highly Qualified Teacher," *Harvard Educational Review* 76 (Winter, 2006), 705.
81. Lee Benson, Ira Harkavy, and John Puckett, *Dewey's Dream: Universities and Democracies in an Age of Education Reform* (Philadelphia: Temple University Press, 2007), 33–44.
82. John Dewey, "The School as Social Centre," in *John Dewey: The Middle Works*, 2: 82.

83. Ibid., 90.
84. I have developed this interpretation of Dewey's views on curriculum and their link to issues of community from the following sources: *Democracy and Education*, 194–206; *The School and Society*, 13–16; John Dewey, "My Pedagogic Creed," in John Dewey: *The Early Works*, 5: 84–95; John Dewey, "The Child and the Curriculum," in *John Dewey: The Middle Works*, 2: 271–291; John Dewey, *Experience and Education* (New York: Collier Books, 1963), 73–88; Katherine Camp Mayhew and Anna Camp Edward, *The Dewey School* (New York: Atherton Press, 1966), 20–36, 336–346; Herbert M. Kliebard, *The Struggle for the American Curriculum*, 3rd ed. (New York: RoutledgeFalmer, 2004), 51–75; Arthur G. Wirth, *John Dewey as Educator: His Design for Work in Education (1894–1902)* (New York: John Wiley & Sons, 1966), 121–134, 161–171.
85. Michael C. Johanek and John Puckett, *Leonard Covello and the Making of Benjamin Franklin High School: Education as if Citizenship Mattered* (Philadelphia: Temple University Press, 2007), 21–47; William Reese, *Power and the Promise of School Reform: Grassroots Movement During the Progressive Era* rev. ed. (New York: Teachers College Press, 2002).
86. Benson et al., xiii, 15–16, 61, 77–78.
87. Ibid., 63–73; Johanek and Puckett, 10–17, 110–120; Elsie Clapp, *Community Schools in Action* (New York: Viking Press, 1939, 66–124 and Dewey's foreword to the volume on pp. vii–x.
88. Leonard Covello, "The School as the Center of Community Life in an Immigrant Area," in *The Community School*, ed. Samuel Everett (New York: D. Appleton-Century, 1938), 128.
89. Johanek and Puckett, 122–132.
90. Ibid., 132–137.
91. Barry M. Franklin, "Education for an Urban America: Ralph Tyler and the Curriculum Field," in *International Perspectives in Curriculum History*, ed. Ivor Goodson (New York: Routledge, 1988), 286–287.
92. Judith Rodin, *The University and Urban Renewal: Out of the Ivory Tower and Into the Streets* (Philadelphia: University of Pennsylvania Press, 2007), 13–22; David J. Maurrasse, *Beyond the Campus: How Colleges and Universities Form Partnerships with their Communities* (New York: Routledge, 2001), 29–64.
93. Rodin, 139–166.
94. Benson, Harkavy, and Puckett, 89–96, 104.
95. Eddie S. Glaude, Jr., *In a Shade of Blue: Pragmatism and the Politics of Black America* (Chicago: University of Chicago Press, 2007), 111–132.
96. Ibid., 132–150.
97. I use the term ethno-historical to refer to studies that explore contemporary issues using ethnographic methods, particularly interviews and focus groups, but then situate those studies in a historical context. These two types of methods have conventions regarding the identification of informants that I honored in the book. The historical chapters were drawn from newspaper accounts and other public documents and involve individuals who are identified in these sources. In those chapters, I saw no reason to provide them

with confidentiality. The ethno-historical chapters explored contemporary events whose participants are still living and professionally active. Where my data was drawn from public documents that identified these informants by name, I did also. Where my information came from interviews, I did not identify my informants by name and in some instances as necessary used pseudonyms to refer to the schools and locations in which they worked.

98. Mitchell Dean, *Critical and Effective Histories: Foucault's Methods and Historical Sociology* (London: Routledge, 1994), 35–36. For a discussion of Foucault's notion of genealogy see Mark Olssen, John Codd, and Anne-Marie O'Neil, *Education Policy, Globalization, Citizenship and Democracy* (London: Sage Publications, 2004), 42–58; Steven Best, *The Politics of Historical Vision: Marx, Foucault, Habermas* (London: Guilford Press, 1999), 110–114; Brent Davis, *Inventions of Teaching: A Genealogy* (Mahwah: Lawrence Erlbaum Associates, 2004),3–5.
99. Michel Foucault, *Discipline and Punish: The Birth of the Prison*, trans. Alan Sheridan (New York: Vintage Books, 1979), 30–31.
100. Michael S. Roth, "Foucault's History of the Present," *History and Theory* 20 (February, 1981), 32–46; Hubert L. Dreyfus and Paul Rabinow, *Michael Foucault: Beyond Structuralism and Hermeneutics*, 2nd ed. (Chicago: University of Chicago Press, 1983), 104–125; Thomas S. Popkewitz, "Curriculum: The History of," *Encyclopedia of Curriculum Studies*, ed. Craig Kridel (Thousand Oaks: Sage Publications, in press).
101. Michel Foucault, "Nietzsche, Genealogy, History," in *Language, Counter-Memory, Practice: Selected Essays and Interviews*, ed. Donald F. Bouchard, trans. Donald F. Bouchard and Sherry Simon (Ithaca: Cornell University Press, 1977), 139–164; Michel Foucault, "Truth and Power," in *Power/Knowledge: Selected Interviews and Other Writings, 1972–1977*, ed. Colin Gordon, trans. Colon Gordon, Leo Marshall, John Mepham, and Kate Sopher (New York: Pantheon Books, 1977), 109–133.
102. Fisher, 167–169.
103. For an example of the depiction of policy outcomes as narratives see John Rogers' characterization of the policy narratives that the Bush administration developed in its promotion of the No Child Left Behind Act. See John Rogers, "Forces of Accountability? The Power of Poor Parents in NCLB," *Harvard Educational Review* 76 (Winter, 2006), 611–641.

2 Community Conflict and Compensatory Education in New York City: More Effective Schools and the Clinic for Learning

1. David K. Cohen, "Teachers Want What Children Need," 1967, 1–2, The Joint Center for Urban Studies, Box 4, Folder 27P, United Federation of Teachers Collection, More Effective Schools Program, Robert F. Wagner Labor Archives, New York University (hereafter MES).

2. Ibid., 2–3.
3. Ibid., 13–14.
4. Alan A. Altshuler, *Community Control: The Black Demand for Participation in Large American Cities* (New York: Pegasus, 1970), 64–65. Not everyone, however, used these terms synonymously. Mario Fantini, Marilyn Gittell, and Richard Magat noted an important difference between these two arrangements. As they saw it, decentralization represented a reorganization of a large school system by dividing it into smaller administrative units representing fewer schools serving smaller geographic areas. Community control, on the other hand, involved the devolution of a degree of authority for governing a school system to its immediate clientele, namely the parents of the children who attend and those adult citizens residing in the geographic area that the school system serves. In effect, then, decentralization could occur without community control, but community control did by its nature entail decentralization of some sort. See Mario Fantini, Marilyn Gittell, and Richard Magat, *Community Control and the Urban School* (New York: Praeger, 1970), 90–99; and Marilyn Gittell and Alan G. Hevesi, *The Politics of Urban Education* (New York: Praeger, 1969), 8–12.
5. Cohen., 15–21.
6. Ibid., 22–23.
7. Richard D. Kahlenberg, *Tough Liberal: Albert Shanker and the Battle Over Schools, Unions, Race, and Democracy* (New York: Columbia University Press, 2007), 57–58, 80–87.
8. Rose Shapiro to Members of the Board of Education, November 28, 1966, Rose Shapiro Papers, Box 7, Folder, Special Collections, Milbank Memorial Library, Teachers College, Columbia University (hereafter RS).
9. Harvey Kantor and Barbara Brenzel, "Urban Education and the 'Truly Disadvantaged': The Historical Roots of the Contemporary Crisis, 1945–1990," in *The Underclass Debate: Views from History*, ed. Michael Katz (Princeton: Princeton University Press, 1992), 366–402; John Rury, "The Changing Social Context of Urban Education: A National Perspective," in *Seeds of Crisis: Public Schooling in Milwaukee since 1920*, ed. John L. Rury and Frank A. Cassell (Madison: University of Wisconsin Press, 1993), 10–41.
10. For various interpretations of this controversy see the following: Derek Edgell, *The Movement for Community Control of New York's City Schools, 1966–1970: Class Wars* (Lewiston: Edwin Mellen Press, 1998); Jerald Podair, *The Strike that Changed New York: Blacks, Whites, and the Ocean Hill-Brownsville Crisis* (New Haven: Yale University Press, 2002); Daniel H. Perlstein, *Justice, Justice: School Politics and the Eclipse of Liberalism* (New York: Peter Lang, 2004); Vincent J. Cannato, *The Ungovernable City: John Lindsay and His Struggle to Save New York* (New York: Basic Books, 2001).
11. Alice O'Conner, "Community Action, Urban Reform, and the Fight Against Poverty: The Ford Foundation's Gray Areas Program," *Journal of Social History* 22 (July, 1996), 586–625; Alice O'Conner, *Poverty Knowledge: Social Science, Social Policy, and the Poor in Twentieth-Century U.S. History* (Princeton: Princeton University Press, 2001), 131–135; Paul Ylvisaker, "Community

Action: A Response to Some Unfinished Business," in *Conscience and Community: The Legacy of Paul Ylvisaker,* ed. Virginia M. Esposito (New York: Peter Lang, 1999), 12–24.

12. *Detroit News,* October 10, 1959, Norman Drachler Papers, Box 41, The Paul and Jean Hanna Archival Collection, The Hoover Institution of War, Revolution and Peace, Stanford University; Harold Silver and Pamela Silver, *An Educational War on Poverty: American and British Policy-making 1960–1980* (Cambridge: Cambridge University Press, 1991), 49–47.
13. Barry M. Franklin, *From "Backwardness" to "At-Risk": Childhood Learning Problems and the Contradictions of School Reform* (Albany: State University of New York, 1994), 1–21; Barry M. Franklin, "A Historical Perspective on Teaching Low-Achieving Children: A First Account," in *Curriculum and Consequence: Herbert M. Kliebard and the Promise of Schooling,* ed. Barry M. Franklin (New York: Teachers College Press, 2000), 128–152.
14. United Federation of Teachers, "Critique of Acting Superintendent's Budget," 1963–64, 4, Box 2, Folder 39, MES.
15. Ad Hoc Committee for Effective Education, "Recommendation of Committee for Effective Schools in Proposed Tutorial Program," October 28, 1963, Box 1, Folder 3, MES; Simon Beagle, "The More Effective Schools Program: The Birth of a New Design in Education, 5–7, Box 1, Folder 25, MES; *United Teacher,* December 12, 1963, Box 1, United Federation of Teachers Collection, Robert F. Wagner Labor Archives, New York University (hereafter UFT).
16. Ad Hoc Committee for Effective Education, *Minutes,* December 22, 1962, January 12, 1963, MES; *United Teacher,* April, 1963, UFT.
17. "UFT Statement on School Integration," *United Teacher Supplement,* December 16, 1963, 8–10, UFT; Diane Ravitch, *The Great School Wars—New York City, 1805–1973: A History of the Public Schools as Battlefield of Social Change* (New York: Basic Books, 1974), 251–266.
18. Ad Hoc Committee for Effective Education, *Minutes,* March 23, 1963, March 26, 1963, May 11, 1963, June 7, 1963, November 16, 1963, March 21, 1964, April 4, 1964, May 7, 1964, June 18, 1964, Box 1, Folder 3, MES.
19. Ad Hoc Committee for Effective Education, *Minutes,* June 4, 1963, Box 1, Folder 3, MES.
20. *United Teacher,* February 2, 1964, UFT.
21. Beagle, 1,7, Box 1, Folder 25, MES; New York City Board of Education, Office of Education Information and Public Relations, "Press Memorandum," Box 2, Folder 43, MES; New York City Board of Education, News Bureau, Office of Education Information Services and Public Relations, "Press Release," June 30, 1964, Box 5, Folder 58, RS.
22. *United Teacher,* June 4, 1964, UFT; Simon Beagle, "Remarks on 'The New York More Effective Schools Program'," Box 1, Folder 25, MES.
23. Podair, 56–57; Perlstein, 25.
24. Donovan to Board of Education, November 3, 1965, Box 1, Folder 8, MES.
25. Center for Urban Education, "More Effective Schools Program," August 31, 1966, 3–9, 13–22, Box 1, Folder 6, MES.

26. Ibid., 8.
27. Ibid., 10–11.
28. New York City Board of Education, "Evaluation of the More Effective Schools Program: Summary Report, September, 1966, MES Evaluations Folder, MES.
29. Nelson Aldrich, "The Controversy over the More Effective Schools: A Special Supplement," *The Urban Review* 2 (May, 1968), 16, Box 4, Folder 15, MES.
30. Shanker, Al. Telephone interview, August 23, 1995.
31. Franklin to Shanker, March 23, 1967, Box 1, Folder 15, MES; Beagle to World Herald Tribune School Page Editor, February 1, 1967, Box 1, Folder 26, MES; Patricia Sexton *An Assessment of the All Day Neighborhood School Program for Culturally Deprived Children* (Cooperative Research Project No. 1527), 1965; Sol Cohen, *Progressives and Urban School Reform: The Public Education Association of New York City, 1895–1954* (New York: Teachers College, Bureau of Publications, 1963).
32. Giardino to Beagle, March 15, 1967, Box 2, Folder 38, MES.
33. Schwager to Giardino, April 7, 1967, Box 2, Folder 38, MES.
34. Brown to Schwager, April 12, 1967, Box 2, Folder 38, MES.
35. Schwager to Lindsay, April 4, 1967, Box 1, Folder 7, MES.
36. Lombard to Lindsay, June 16, 1967, Box 3, Folder 57, MES; Lindsay to Students, June 26, 1967, Box 3, Folder 57, MES.
37. Lomard and Broakstone to Lindsay, June 30, 1967, Box 3, Folder 57, MES.
38. Brown to Shanker, May 10, 1967, Box 1, Folder 9, MES.
39. Schwager to Hall, May 19, 1967, Box 1, Folder 7, MES.
40. Cogan to O'Daly, July 14, 1967, Box 1, Folder 16, MES.
41. New York City Council Resolution 999, July 11, 1967, Box 1, Folder 7, MES.
42. Aldrich, 16; "A Statement by the 'Joint Planning Committee for More Effective Schools' on the Superintendent's Directive to Decentralize the Administration of the More Effective Schools Program, and to Reduce More Effective School Staffs by Eliminating the Positions of the Audio-Visual Teacher and the Health Counselor," June 13, 1967, Box 2, Folder 43, MES.
43. Siegel to Shanker, June 9, 1967, Box 2, Folder 38, MES.
44. Podair, 90–91.
45. Stephen Cole, *The Unionization of Teachers: A Case Study of the UFT* (New York: Praeger, 1969), 188–191.
46. Marjorie Murphy, *Blackboard Unions: The AFT and the NEA, 1900–1980* (Ithaca: Cornell University Press, 1990), 237.
47. David J. Fox, "Expansion of the More Effective School Program," Center for Urban Education, September, 1967, 3–4, MES Evaluation Folder, MES.
48. Ibid., 120–122, MES.
49. Sidney Schwager, "An Analysis of the Evaluation of the MES Program Conducted by the Center for Urban Education," *The Urban Review* 2 (May, 1968), 18–23, Box 4, Folder 15, MES.

50. Sidney Schwager, "More Effective Schools Program—An Analysis of the Latest Evaluation by the Center for Urban Education." An Address to the Delegates of the AFT 1967 Convention, Box 1, Folder 8, MES.
51. Schwager to Kivelson, October 25, 1967, Box 1, Folder 26, MES.
52. Schwager to Turkel, November 28, 1967, Box 1, Folder 26, MES.
53. Donovan to Members of the Board of Education, November 30, 1967, Box 2, Folder 39, MES.
54. MES Parents, n.d., Box 2, Folder 42, MES.
55. New York City Schools, "Report of Joint Planning Committee for More Effective Schools to the Superintendent of Schools," May 15, 1964, i, ii, 18, Box 4, Folder 26P, MES.
56. The Psychological Corporation, "Evaluation Report for the Project More Effective Schools Program in Poverty Area Schools, 1968–69," 150, 160, Box 4, Folder 9P, MES.
57. Joseph Alsop, "No More Nonsense About Ghetto Education!" *New Republic* 157 (July 22, 1967), 18–20.
58. Ibid., 20–21
59. Robert Schwartz, Thomas Pettigrew, and Marshall Smith, "Fake Panaceas for Ghetto Education: A Reply to Joseph Alsop," *New Republic* 157 (September 23, 1967), 16–19.
60. The Psychological Corporation, 12–15.
61. Ibid., 139.
62. National Council for Effective Schools, "Minutes of the Mid West Regional MES Conference, November 16, 1968, American Federation of Teachers Collection, Archives of Labor History and Urban Affairs, Walter P. Reuther Library, Wayne State University.
63. Pearson to Beagle October 23, 1966, Box 2, Folder 35, MES; Linne to Schwager, November 18, 1966, Box 2, Folder 35, MES; Rothernel to Schwager, August 14, 1967, Box 2, Folder 35, MES; Cybulski to Beagle, September 27, 1967, Box 2, Folder 35, MES.
64. Si Beagle, "The AFT's 1967–68 MES Campaign-A Tentative Report," January 15, 1968, Box 1, Folder 25, MES.
65. Schwartz to Beagle, October 7, 1966, Box 1, Folder 7, MES; *New York Times*, July 17, 1966.
66. Robertson to Donovan, March 3, 1966, Bernard Donovan Papers, Box 8, Folder 2, Milbank Memorial Library, Teachers College, Columbia University (hereafter BDP); Donovan to Fantini, May 4, 1966, Box 8, Folder 2, BDP; Robertson to Donovan, May 11, 1966, BDP; Donovan to Fantini, June 17, 1966, BDP.
67. "Agreement between New York University and the New York City Board of Education, September 9, 1966, BDP.
68. New York University Capital Program, "Education-Problems in Bedford-Stuyvesant," 1–25, Box 8, Folder 2, BDP.
69. Ibid, Appendix B.
70. John C. Robertson, "The Clinic for Learning-Elements for Replication," 3–4, Box 8, Folder 2, BDP.

71. John C. Robertson and Neil M. Postman, "Clinic for Learning," 2–3, Box 8, Folder 2, BDP.
72. Gerry Rosenfield, *A Study of the New York University Clinic for Learning* (New York: Center for Urban Education, 1970), 12.
73. Robertson, 4–7, Box 8, Folder 2, BDP.
74. Robertson to Donovan, October 24, 1966, Box 8, Folder 2, BDP.
75. Robertson to Donovan, November 9, 1966, Box 8, Folder 2, BDP.
76. Robertson to Donovan, December 5, 1966, Box 8, Folder 2, BDP.
77. Robertson to Donovan, February 1, 1967, Box 8, Folder 2, BDP.
78. Schwartz to Donovan, February 9, 1967, Box 8, Folder 2, BDP.
79. Rosenfield, 14–15.
80. Ibid., 22–23.
81. Ibid., 17–18.
82. Ibid., 26–28.
83. Ibid., 21–22.
84. Donovan to Griffiths, May 9, 1967, Box 8, Folder 2, BDP.
85. Rosenfield, 16, BDP.
86. Robertson to Donovan, April 18, 1967, Box 8, Folder 2, BDP.
87. "A Report of the Community Committee for the Evaluation of the Clinic for Learning and Junior High 57-K, 1–2, Box 8, Folder 2, BDP.
88. Ibid., 2–3.
89. Ibid., 3–7.
90. Schwartz to Donovan, January 26, 1968, Box 8, Folder 2, BDP.
91. Clarke to Donovan, January 31, 1968, Box 8, Folder 2, BDP.
92. Robertson to Donovan, February 9, 1968.
93. Podair, 34–36; 159–160; Cannato, 272–275; East Harlem Protestant Parish, *Newsletter*, October 1966, Box 3, Folder 18, RS; "A Summary of the Controversy at I.S. 201, "*IRCD Bulletin*, II and III (Winter, 1966–67), 1–2, Box 2, Folder 65, UFT; East Harlem Protestant Parish, *News Letter*, October, 1966, Box 3, Folder 18, RS.
94. New York City Board of Education, "Proposals for Improving Education in Schools in Disadvantaged Areas," October 20, 1966, Box 3, Folder 18, RS.
95. Parent/Community Negotiating Committee, Intermediate School 201, "Response to the Board of Education 'Proposals for Improving Education in Schools in Disadvantaged Areas'," 1–4, Box 3, Folder 18, RS.
96. "The Institute of Learning (for the 'Disruptive Child')," n.d., 5, Box 10, Folder 7, BDP.
97. Ibid, 16–17, BDP; Malina Imiri Abudadika (Sonny Carson), *The Education of Sony Carson* (New York: W.W. Norton, 1972), 168–174.
98. Donovan to Robert Carson, January 31, 1968, Box 10, Folder 7 BDP.
99. Herbert to Donovan, February 2, 1968, Box 10, Folder 7, BDP.
100. Donavan to Herbert, February 7, 1968, Box 10, Folder 7, BDP.
101. Brooklyn and Bronx CORE, February 9, 1968, Box 10, Folder 7, BDP.
102. Abudadika, 134.
103. Ibid., 140.
104. Ibid., 203.

105. James T. Patterson, *Brown v. Board of Education: A Civil Rights milestone and Its Troubled Legacy* (New York: Oxford University Press, 2001), 70–78; William J. Reese, *America's Public Schools: From the Common School to "No Child Left Behind"* (Baltimore: Johns Hopkins University Press, 2005), 226–230; Charles Vert Willie and Sarah Susannah Willie, "Black, White, and Brown: The Transformation of Public Education in America," *Teachers College Record* 107 (March, 2005), 475–495; Clayborn Carson, "Two Cheers for Brown v. Board of Education," *The Journal of American History* 91 (June, 2004), 26–31.
106. Ralph Ellison to M.D.Sprague, May 19, 1954 in John F. Callahan, "American Culture Is of a Whole: From the Letters of Ralph Ellison,"*New Republic* 220 (March 1, 1999), 34–49.
107. Richard Kluger, *Simple Justice* (New York: Alfred A. Knopf, 1976), 710
108. William L. Van Deburg, *New Day in Babylon: The Black Power Movement and American Culture, 1965–1975* (Chicago: University of Chicago Press, 1992), 20–53, 113–131; William C. Dawson, *Black Visions: The Roots of Contemporary African-American Political Ideologies* (Chicago: University of Chicago Press, 2001), 85–134.

3 Community, Race, and Curriculum in Detroit: The Northern High School Walkout

1. *Detroit* Free *Press*, April 8, 1966.
2. David Gracie, "The walkout at Northern High," *New University Thought* 5 (1967), 13–28; Karl Gregory, "The walkout: Symptom of Dying Inner City Schools," New *University Thought* 5 (1967), 29–54.
3. National Commission on Professional Rights and Responsibilities, *Detroit Michigan: A Study of Barriers to Equal Educational Opportunity in a Large City* (Washington, DC: National Education Association, 1967), 73–84.
4. Sidney Fine, *Violence in the Model City: The Cavanagh Administration, Race Relations, and the Detroit Riot of 1967* (Ann Arbor: University of Michigan Press, 1989), 52–57.
5. Jeffrey Mirel, *The Rise and Fall of an Urban School System: Detroit, 1907–81* (Ann Arbor: University of Michigan Press, 1992), 298–313.
6. Robert Booth Fowler, "Community: Reflections on Definition," in *New Communitarian Thinking: Person, Virtues, Institutions, and Communities*, ed. Amitai Etzioni (Charlottesville: University Press of Virginia, 1995), 89–90; Robert Booth Fowler, *The Dance with Community: The Contemporary Debate in American Political Thought* (Lawrence: University Press of Kansas, 1991), 103–118.
7. Gracie, 16–17; *Detroit News*, April 13, 1966, Detroit Public School Archives (hereafter DPSA); *Detroit News*, April 18, 1966, Norman Drachler Papers, Box 41, The Paul and Jean Hanna Archival Collection, The Hoover

Institution of War, Revolution and Peace, Stanford University (hereafter NDP); *Detroit Free Press*, April 19, 1966, Box 41, NDP.

8. Gracie, 17; Detroit Board of Education, *Proceedings of the Board of Education of the City of Detroit*, April 12, 1965, 505–508, DPSA; *Detroit News*, April 13, 1966, DPSA; *Detroit Free Press*, April 18, 1966, Box 41, NDP.
9. National Commission, 77–78; *Proceedings of the Board of Education of the City of Detroit*, April 26, 1966, 526, DPSA.
10. Gracie, 18–20; *Detroit Free Press*, April 20, 1966, DPSA; *Detroit News*, April 21, 1966, Box 41, NDP.
11. Fine, 50–53; Gracie, 18–24; Mirel, 300–304; *Detroit Free Press*, April 18, 1966, Box 41, NDP; *Detroit News*, April 21, 1966, Box 41, NDP.
12. National Commission, 11.
13. *Detroit Free Press*, April 28, 1966, Box 41, NDP.
14. Mirel, 193–194, 258–259.
15. I have drawn my account of these demographic and economic changes from several sources. Most important is Thomas J. Sugrue's *The Origin of the Urban Crisis: Race and Inequality in Postwar Detroit* (Princeton: Princeton University Press, 1996), Chapters 2, 4–6. Other sources that proved helpful include Fine, Chapters 1–7; Peniel E. Joseph, *Waiting 'Til the Midnight Hour: A Narrative History of Black Power in America* (New York: Owl Books, 2006), 52–67. Joe T. Darden, Richard Child Hill, June Thomas, and Richard Thomas, *Detroit: Race and Uneven Development* (Philadelphia: Temple University Press, 1987), Chapters 1–3, 5; Albert J. Mayer and Thomas F. Hoult, *Race and Residence in Detroit* (Detroit: Institute for Urban Studies, Wayne State University, 1962), Box 15, NDP.
16. Sugrue, 23.
17. *Detroit Free Press*, April 27, 1966, DPSA.
18. *Detroit News*, April 19, 1966, Box 41, NDP.
19. *Detroit Free Press*, April 24, 1966, Box 41, NDP.
20. *Report of Community Hearings*, May, 1966, National Association for the Advancement of Colored People-Detroit Branch Papers, Part II, Box 12, Folder 17, Archives of Labor History and Urban Affairs, Walter P. Reuther Library, Wayne State University (hereafter NAACP-II).
21. William. Ardrey, "A Statement Concerning the Northern High School Situation," April 22, 1966, 1, Box 12, Folder 17, NAACP-II.
22. Gracie, 14; *Detroit Free Press*, April 8, 1966, April 20, 1966, DPSA; *Detroit Free Press*, May 1, 1966, Box 41, NDP.
23. *Detroit News*, April 19, 1966, April 25, 1966, Box 41, NDP.
24. Charles. Lewis, "What Makes Sammy Fail," n.d., Remus G. Robinson Papers, Box 6, Folder 27, Archives of Labor History and Urban Affairs, Walter P. Reuther Library, Wayne State University (hereafter RRP).
25. Mirel, 239–240.
26. Detroit Board of Education, *Consultants' Report to the School Program (Curriculum) Sub-Committee of the Citizens Advisory Committee on School Needs*, 1958, 27, DPSA.

27. Ibid., 93–100, DPSA; Detroit Board of Education, *Citizens Advisory Committee on School Needs: Findings and Recommendations (abridged)*, 1958, 22, 40, DPSA.
28. Detroit Board of Education, *One Year After: A Staff Report on What Has Transpired since the Citizens Advisory Committee on School Needs Made Public Its Report and Recommendations for the Decade Ahead*, December 7, 1959, DPSA.
29. Mirel, 259–264; *Detroit Free Press*, January 13, 1960, January 26, 1960, February 28, 1960, DPSA.
30. Detroit Board of Education, *Findings and Recommendations of the Citizens Advisory Committee on Equal Educational Opportunities (abridged)*, March, 1962, 4, DPSA.
31. Ibid., 15, DPSA.
32. Ibid., 18–20, 22, DPSA.
33. Detroit Board of Education, *Detroit's Citizens Advisory Committee on School Needs Supplementary Materials*, Part I, 1957–58, 2–3, DPSA.
34. Ibid., 6–7, DPSA.
35. Ibid., 23, 32, DPSA.
36. Ibid., 21, DPSA.
37. Ibid., 16, 21, DPSA.
38. *Findings and Recommendation of the Citizens Advisory Committee on Equal Educational Opportunity*, 57–62.
39. *Detroit Free Press*, March 12, 1962, Box 9, NDP.
40. Ibid.
41. *Detroit Free Press*, June 21, 1960, DPSA.
42. Ibid., DPSA.
43. Citizens Advisory Committee on Equal Educational Opportunities, *Minutes*, September, 1960, 73–74, DPSA.
44. Ibid., 74, DPSA.
45. For a discussion of this viewpoint as it applies to Detroit See Sugrue, Chapters 7–9; Thomas Sugrue, "Crabgrass-Roots Politics: Race, Rights, and the Reaction Against Liberalism in the Urban North, 1940–1964," *Journal of American History* 82 (1995), 551–578; Heather. Thompson, *Whose Detroit: Politics, Labor, and Race in a Modern American City* (Ithaca: Cornell University Press, 2001), Chapters 1–4. Arnold Hirsch offers a similar perspective for racial relations in Chicago. See Arnold Hirsch, "Massive Resistance in the Urban North: Trumbull Park, Chicago, 1953–1966," Journal *of American History* 82 (1995), 522–550; Arnold. Hirsch, *Making the Second Ghetto: Race and Housing in Chicago, 1940–1960* (Cambridge: Cambridge University Press, 1983), Chapters 3, 6. A broader and more national discussion of this interpretative framework is provided by Gary Gerstle. See Gary Gerstle, "Race and the Myth of the Liberal Consensus," *Journal of American History* 82 (1995), 579–586; Gary Gerstle, *American Crucible: Race and Nation in the Twentieth Century* (Princeton: Princeton University Press, 2001), Chapter 7.
46. C.R.F. to Robinson, n.d., Box 5, Folder 9, RRP.

47. White UM female student to Robinson, February 2, 1959, Box 16, Folder 27, RRP.
48. A Northwestern Teacher, "What's Wrong with Our Schools," *The Illustrated News*, February 12, 1962, Box 7, Folder 55, RRP.
49. Peter Eisinger, *Patterns of Interracial Politics: Conflicts and Cooperation in the City* (New York: Academic Press, 1978), 2–8; James Jennings, *The Politics of Black Empowerment: The Transformation of Black Activism in America* (Detroit: Wayne State University Press, 1992), 123–124; Steven Gregory, *Black Corona: Race and the Politics of Place in an Urban Community* (Princeton University Press, 1998), 11–13; Michael Omi and Howard Winant, *Racial Formation in the United States from the 1960s to the 1990s*, 2nd ed. (New York: Routledge, 1994), 99–101.
50. David Katzman, *Before the Ghetto: Black Detroit in the Nineteenth Century* (Urbana: University of Illinois Press, 1973), 81–103, 163.
51. August Meier and Elliott Rudwick, *Black Detroit and the Rise of the UAW* (New York: Oxford University Press, 1979), 10–20, 33.
52. "Northeastern, May 11, 1966," Detroit Urban League Collection, Box 54, Education and Youth Incentives Folder, Bentley Historical Library, University of Michigan (hereafter DUL).
53. *Michigan Chronicle*, October 1, 1966.
54. Mirel, 299; "Open letter to the Detroit Board of Education and to the Public from the Ad Hoc Executive Committee," Detroit Commission on Community Relations Collection, Human Rights Department, Part III, Box 49, Folder 8, Archives of Labor History and Urban Affairs, Walter P. Reuther Library, Wayne State University (hereafter DCCR).
55. Wells statement, Box 45, Folder 3, DCCR.
56. "Massive Underachievement in the Detroit Public Schools: A Statement by the Ad Hoc Committee of Citizens for Equal Educational Opportunity made to the Detroit Board of Education, September 1, 1968," Box 49, Folder 6, DCCR.
57. "Presentation of June Shagaloff to the Detroit High School Study Commission," May 31, 1967, 10, Box 68, Reference File-Education (1) Folder, DUL.
58. Ibid., 5
59. Ibid., 4–5.
60. Michael Dawson, *Black Visions: The Roots of Contemporary African-American Political Ideologies* (Chicago: University of Chicago Press, 2001), 21–23; Richard W. Thomas, *Life for Us Is What We Make It: Building Black Community in Detroit, 1915–1945* (Bloomington: Indiana University Press, 1992), 194–201; Judith Stein, *The World of Marcus Garvey: Race and Class in Modern Society* (Baton Rouge: Louisiana State University Press, 1986), 230–232; Claude Andrew Clegg III, *An Original Man: The Life and Times of Elijah Muhammad* (New York: St. Martin's Press, 1997), 14–37.
61. Ibid., 85–102; Jennings, 95–101, 116–118; Dan Georgakas and Marvin Surkin, *Detroit: I Do Mind Dying. A Study in Urban Revolution* (New York: St. Martin's Pres, 1975), 1–2, 91–93; Wilbur Rich, *Coleman Young and Detroit Politics: From Social Activist to Power Broker* (Detroit: Wayne State University Press, 1989), 82–83, 273–275; Heather Thompson, "Rethinking the Collapse

of Postwar Liberalism: The Rise of Mayor Coleman Young and the Politics of Race in Detroit," in *African American Mayors: Race, Politics, and the American City*, ed. David Colburn and Jeffrey Adler (Urbana: University of Illinois Press, 2001), 230.

62. *Michigan Chronicle*, June 18, 1966, June 25, 1966, August 6, 1966.
63. *Michigan Chronicle*, June 4, 1966.
64. *Michigan Chronicle*, May 28, 1966.
65. Ibid.
66. *Michigan Chronicle*, October 28, 1967.
67. *Michigan Chronicle*, November 4, 1967.
68. *Michigan Chronicle*, November 23, 1968, November 30, 1968, December 14, 1968.
69. *Michigan Chronicle*, March 16, 1968.
70. Black Ministers-Teachers Conference, "Declaration of Black Teachers," April 27, 1968, Papers of New Detroit, Inc., Box 47, Folder 7, Archives of Labor History and Urban Affairs, Walter P. Reuther Library, Wayne State University.
71. *Detroit Free Press*, October 1, 1969, Box 46, Folder 6, DCCR; Northern's Black student demands, n.d., Box 46, Folder 6, DCCR.
72. Inner City Parents Council, "Program for Quality Education in Inner City Schools," June 13, 1967, 4, Box 29, Folder 2, NAACP-II.
73. Ibid, 12.
74. Ibid., 11.
75. *Detroit Free Press*, April 20, 1966, Box 41, NDP.
76. Report of Community Hearings, NAACP-II.
77. Post Junior High School, "Sub-Committee Investigating the Concerns of the Community and Recommendations Made to the School and the Board of Education Previous to the Walkout," May 1, 1968, Box 45, Folder 35, DCCR.
78. Jerald Podair, *The Strike that Changed New York: Blacks, Whites, and the Ocean Hill-Brownsville Crisis* (New Haven: Yale University Press, 2002), p. 5.
79. For a discussion of this understanding of community see Gregory, 10–11.
80. Cornell West, *The Cornell West Reader* (New York: Basic Books, 1999), 591. For a discussion of the interplay between notions of identity and race see Eddie S. Glaude, Jr. *In a Shade of Blue: Pragmatism and the Politics of Black America* (Chicago: University of Chicago Press 2007), 47–65; Janet E. Helms, "An Overview of Black Racial Identity Theory," in *Black and White Racial Identity*, ed. Jane E. Helms (New York: Greenwood Press, 1990), 9–32.
81. Elizabeth Frazer, *The Problems of Communitarian Politics: Unity and Conflict* (New York: Oxford University Press, 1999), 83.
82. Robin. Kelly, *Race Rebels: Culture, Politics, and the Black Working Class* (New York: Free Press, 1994), 38–39.
83. West, 502.
84. Mirel, 293–409; Jeffrey Mirel, "After the Fall: Continuity and Change in Detroit, 1981–1995," *History of Education Quarterly* 38 (Autumn, 1998), 237–267.

4 RACE AND COMMUNITY IN A BLACK LED CITY: THE CASE OF DETROIT AND THE MAYORAL TAKEOVER OF THE BOARD OF EDUCATION

1. John Engler, "Michigan the 'Smart State': First in the 21st Century," 1999 State of the State Address, retrieved from www.migov.state.mi.us/speeches/sos1999.html.
2. *Detroit News*, March 21, 1999; *New York Times*, June 29, 1995; Anthony S. Byrk, David Kerbow, and Sharon Rollow, "Chicago School Reform," in *New Schools for a New Century: The Redesign of Urban Education*, ed. Diane Ravitch and Joseph R. Viteritti (New Haven: Yale University Press, 1997), 164–200; Stefanie Chambers, *Mayors and Schools: Minority Voices and Democratic Tensions in Urban Education* (Philadelphia: Temple University Press, 2006), 25–26. Peter Eisinger, "Mayoral Takeover in Detroit," (paper presented at the annual meeting of the American Educational Research Association, New Orleans, La., April, 2000); Jeffrey R. Henig and Wilbur C. Rich, "Mayor-centrism in Context," in *Mayors in the Middle: Politics, Race, and Mayoral Control of Urban Schools*, ed. Jeffrey R. Henig and Wilbur C. Rich (Princeton: Princeton University Press, 2004), 5–7; Jeffrey Mirel, "School Reform Chicago Style: Educational Innovation In a Changing Urban Context," *Urban Education* 28 (July, 1993), 116–149.
3. Jeffrey Mirel, "After the Fall: Continuity and Change in Detroit, 1981–1995," *History of Education Quarterly* 38 (Fall, 1998), 237–267.
4. David Arsen, David Plank, and Gary Sykes, *School Choice Politics in Michigan: The Rules Matter* (East Lansing: Michigan State University, 1999); C. Phillip Kearney, *A Primer on Michigan School Finance*, 3rd ed. (Ann Arbor: University of Michigan, 1994).
5. *Detroit Free Press*, January 25, 1997.
6. *Detroit News*, January 29, 1997, February 2, 1997.
7. In revising and updating this chapter, I benefited immensely from a number of books and articles that were written after I wrote an earlier version of this essay. The best two, although my interpretive framework differs from theirs, are Jeffrey Mirel, "Detroit: 'There is Still a Long Road to Travel, and Success is Far from Assured'," in *Mayors in the Middle: Politics, Race, and Mayoral Control of Urban Schools*, ed. Jeffrey R. Henig and Wilbur C. Rich (Princeton: Princeton University Press, 2004), 120–158; and Wilbur C. Rich, "Whose Afraid of a Mayoral Takeover of Detroit Public Schools," in *When Mayors Take Charge: School Governance in the City*. Ed. Joseph P. Viteritti (Washington, DC: Brookings Institution Press, 2009), 148–167.
8. *Detroit Free Press*, January 1, 1999.
9. *Detroit News*, January 27, 1999.
10. *Detroit Free Press*, January 1, 1999; *Detroit News*, February 21, 1999, February 23, 1999.
11. Michigan Senate, *Senate Bill No. 297*, February 10, 1999.

12. *Detroit News*, February 14, 1999; *Michigan Chronicle*, January 27–February 4, 1999.
13. *Detroit Free Press*, January 27, 1999; *Detroit News*, February 14, 1999.
14. *Detroit News*, February 14, 1999.
15. Ibid.
16. *Detroit News*, February 11, 1999.
17. *Detroit Free Press*, February 13, 1999.
18. *Detroit News*, February 21, 1999.
19. *Detroit News*, February 11, 1999, February 18, 1999.
20. *Detroit News*, February 14, 1999.
21. *Detroit News*, February 10, 1999.
22. *Detroit Free Press*, January 30, 1999.
23. *Michigan Chronicle*, December 23–29, 1998.
24. *Detroit News*, February 11, 1999.
25. Dennis Archer, "The Road to Excellence," State of the City Address (Detroit: City of Detroit Executive Office, 1999).
26. *Detroit News*, February 23, 1999; *Flint Journal*, February 23, 1999.
27. *Detroit Free Press*, February 24, 1999.
28. *Detroit News*, February 17, 1999, February 24, 1999.
29. *Detroit Free Press*, February 15, 1999.
30. Jeffrey R. Henig, Richard C. Hula, Marion Orr, and Desiree S. Pedescleaux, *The Color of School Reform: Race, Politics, and the Challenge of Urban Education* (Princeton: Princeton University Press, 1999), 39.
31. *Detroit News*, February 17, 1999.
32. *Detroit News*, February 18, 1999.
33. *Detroit Free Press*, February 25, 1999.
34. *Flint Journal*, February 22, 1999.
35. *Detroit Free Press*, March 18, 1999; *Detroit News*, March 18, 1999.
36. *Detroit Free Press*, March 26, 1999.
37. Rich, *Whose Afraid*, 154; *Detroit Free Press*, March 19, 1999; *Detroit News*, March 19, 1999.
38. *Detroit News*, March 26, 1999; *Flint Journal*, March 26, 1999.
39. Derek W. Meinecke and David W. Adamany, "School Reform in Detroit and Public Act 10: A Decisive Legislative Effort with an Uncertain Outcome," *The Wayne Law Review* 47 (Spring, 2001), 8–9, retrieved June 12, 2008 from www.lexisnexis.com/us/Inacademic/frame.do?tokenKey=rsh-20.966948.194905731; *Detroit Free Press*, March 26, 1999.
40. *Detroit News*, February 14, 1999; *Detroit Free Press*, February 13, 1999.
41. *Detroit Free Press*, January 27, 1999.
42. *Detroit Free Press*, February 17, 1999.
43. Frederick M. Hess, *Spinning Wheels: The Politics of Urban School Reform* (Washington, DC: Brookings Institution Press, 1999; Joseph Murphy, "Restructuring America's Schools: An Overview," in *Education Reform in the '90s*, ed. Chester E. Finn and Theodor Rebarber (New York: Macmillan, 1992), 3–20.

44. *Detroit Free Press*, January 27, 1999, February 15, 1999; *Detroit News*, February 14, 1999.
45. *Michigan Chronicle*, March 10–16, 1999.
46. *Detroit News*, February 14, 1999, February 25, 1999.
47. *Detroit News*, February 25, 1999.
48. *Detroit News*, February 11, 1999.
49. *Detroit News*, March 21, 1999; *Detroit Free Press*, February 16, 1999.
50. *Detroit Free Press*, February 2, 1999, February 13, 1999, February 18, 1999, February 27, 1999, March 9, 1999.
51. *Detroit Free Press*, February 27, 1999.
52. *Detroit Free Press*, March 11, 1999.
53. *Detroit Free Press*, March 16, 1999; *Detroit News*, February 17, 1999.
54. *Detroit News*, February 14, 1999, March 28, 1999.
55. *Detroit News*, February 17, 1999; *Detroit Free Press*, February 26, 1999.
56. *Detroit Free Press*, March 18, 1999; *Detroit News*, March 19, 1999.
57. *Detroit Free Press*, March 18, 1999.
58. *Detroit News*, February 22, 1999.
59. *Detroit News*, February 25, March 19, 1999.
60. Mirel, "Detroit," 133.
61. Rich, "Whose Afraid of a Mayoral Takeover," 154; *Detroit Free Press*, May 11, 2000.
62. Mirel, "Detroit," 134–138.
63. *Detroit Free Press*, May 5, 2000; *Detroit News*, May 4, 2000, May 5, 2000.
64. *Detroit Free Press*, July 7, 2000.
65. *Detroit Free Press*, June 29, 2001, December 15, 2000; *Detroit News*, July 5, 2001, July 6, 2001.
66. *Detroit Free Press*, November 11, 2000, November 21, 2000, December 15, 2000; *Detroit News*, October 1, 2000, November 20, 2000, November 21, 2000, November 22, 2000, December 4, 2000.
67. Mirel, "Detroit," 142.
68. *Detroit Free Press*, January 7, 2004.
69. Kenneth Burnley, "State of the District Message," February 17, 2003, retrieved June 18, 2008 from www.detroit.k12.mi.us/admin/ceo/pdfs/stateofdistrict2003.pdf; Kenneth Burnley, "State of District Message," May 12, 2004," retrieved June 18, 2008 from www.detroit.k12.mi.us/admin/ceo/pdfs/stateofdistrict2004.pdf.
70. *Detroit News*, October 24, 2004.
71. "Detroit on 'Reform Board: We'll Do It Ourselves Thanks'," *Bridges4Kids*, retrieved December 19, 2004 from www.bridges4kids/articles/4–03/MIRS4–8-03.html.
72. *Detroit News*, October 24, 2004.
73. *Detroit News*, March 11, 2004, April 29, 2004, May 7, 2004, May 13, 2004.
74. *Detroit News*, October 24, 2004.
75. *Michigan Chronicle*, February 11, 2002; *Detroit Free Press*, September 17, 2003; Kwame Kilpatrick, "Education Reform," November 18, 2003, retrieved

December 19, 2004 from www.ci.detroit.mi.us/mayor/speeches/Education%20 Reform%202003.htm.

76. *Detroit Free Press*, January 14, 2004, November 3, 2004.
77. *Detroit Free Press*, October 8, 2004; *Michigan Chronicle*, October 27–November 4, 2004.
78. *Detroit Free Press*, October 22, 2004, October 26, 2004; *Detroit News*, October 26, 2004, October 27, 2004.
79. *Detroit Free Press*, October 8, 2004; *Detroit News*, October 10, 2004, October 27, 2004; *Michigan Chronicle*, October 20–October 26, 2004, October 27–November 2, 2004, November 3–November 9, 2004
80. Barry M. Franklin, "Achievement, Race, and Urban School Reform in Historical Perspective: Three Views from Detroit," *Educational Research and Perspectives* 31 (December, 2004), 24–25.
81. *Detroit Free Press*, November 3, 2004, November 4, 2004; *Detroit News*, November 4, 2004, October 24, 2005; November 8, 2005. See also Chambers, 35–36; Kenneth K. Wong, Francis X. Shen, Dorethea Anagnostopoulos, and Stacey Rutledge, *Improving America's Schools: The Education Mayor* (Washington, DC: Georgetown University Press, 2007), 46–47.
82. Wong et al., 25.
83. David Labaree, *How to Succeed in School without Really Learning* (New Haven: Yale University Press, 1997), 15–52.

5 Educational Partnerships, Urban School Reform, and the Building of Community

1. New York Legislature, *New York State Consolidated Education Law: Article 56 Charter Schools* (December, 1998), 4–7.
2. *New York Times*, March 30, 2001, April 1, 2001, April 3, 2001.
3. Barry M. Franklin, Marianne N. Bloch, and Thomas S. Popkewitz, "Educational Partnerships: An Introductory Framework," in *Educational Partnerships and the State: The Paradoxes of Governing Schools, Children, and Families*, ed. Barry M. Franklin, Marianne N. Bloch, and Thomas S. Popkewitz (New York: Palgrave Macmillan, 2003), 4–5.
4. New York Legislature, 12, 19.
5. *New York Times*, March 20, 2001; New York City Board of Education, "Request for Proposal #1B434-SURR to Charter School Services," August 8, 2000, 1; Diane Ravitch and Joseph P. Viteritti, "Introduction," in *City Schools: Lessons from New York*, ed. Diane Ravitch and Joseph P. Viteritti (Baltimore: Johns Hopkins University Press, 2000), 3–4.
6. Ravitch and Viteritti, "Introduction," 4.
7. Ibid., 5–6.
8. Emanuel Tobier, "Schooling in New York City: The Socioeconomic Context," in Ravitch and Viteritti, 37.
9. Ibid., 26–34, 42.

10. Request for Proposal #1B434, 8–9, 14–15.
11. Association of Community Organization for Reform Now et al. v. New York City Board of Education, Supreme Court of the State of New York, County of Kings, *Verified Petition*, February 27, 2001; Association of Community Organization for Reform Now et al. v. New York City Board of Education, Supreme Court of the State of New York, County of Kings, *Order to Show Cause and Temporary Restraining Order*, February 28, 2001. I obtained these court petitions from the office of ACORNs legal counsel, Arthur Schwartz.
12. The research reported in this section of the chapter involved a telephone interview in April of 2001 with a official of ACORN and in-person interviews in June of 2001 with ACORN's legal counsel, two officials from Edison Schools, and two administrators from the Office of Charter Schools at the New York City Board of Education. Those participating in these interviews were guaranteed anonymity and are thus not identified.
13. Proposal #1B434, 5–10.
14. *New York Times*, March 22, 2001, April 3, 2001.
15. *New York Times*, March 30, 2001.
16. *New York Times*, March 22, 2001.
17. *New York Times*, March 29, 2001.
18. *New York Times*, March 22, 2001.
19. *New York Times*, March 30, 2001.
20. Ibid.
21. *New York Times*, April 3, 2001.
22. *New York Times*, April 7, 2001.
23. *New York Times*, April 7, 2001.
24. Annenberg Institute for School Reform, *Citizens Changing their Schools: A Midterm Report of the Annenberg Challenge*, April, 1999, retrieved August 1, 2004 from http://www.annenberginstitute.org/Challenge/pubs/citizens_changing/contents.html.
25. Ibid.
26. Ibid.; Aida Rodriguez, Joseph A. Pereira, and Shana Brodnax, "Latino Nonprofits: The Role of Intermediaries in Organizational Capacity Building," in *A Future for Everyone: Innovative Social Responsibility and Community Partnerships*, ed. David Maurrasse with Cynthia Jones (New York: Routledge, 2004), 79–100; Meredith I. Honig, "The New Middle Management: Intermediary Organizations in Educational Policy Implementation," *Educational Evaluation and Policy Analysis* 26 (Spring, 2004), 66–68; American Youth Policy Forum, *Local Intermediary Organizations: Connecting the Dots for Children, Youth, and Families*, n.d., retrieved September 19, 2004 from http://www.aypf.org/publications/intermediaries.
27. Annenberg Institute for School Reform, *The Annenberg Challenge: Lessons and Reflections on Public School Reform* (Providence: Annenberg Institute for School Reform, 2002), 32; Robert Rotham, "'Intermediary Organizations' Help Bring Reform to Scale," *Challenge Journal* 6 (Winter, 2002/03), 1–7.
28. NYNSR Research Collaborative, *New York Networks for School Renewal (NYNSR): An Implementation Study* (New York: Institute for Education and

Social Policy, New York University, 2001), 4; Caroline Hendrie, "N.Y.C. Students at Annenberg Sites Were 'Well Served,' Report Finds," *Education Week* (March 6, 2002), 15.

29. Deborah Meier, *The Power of their Ideas: Lessons for America from a Small School in Harlem* (Boston: Beacon Press, 1995), 15–38; Seymour Fliegel with James MacGuire, *Miracle in East Harlem: The Fight for Choice in Public Education* (New York: Time Books, 1993), 31–44; *Center for Collaborative Education* Web site, retrieved May 18, 2001 from http://ww.cce.org.
30. The research reported in this section of the chapter took place in June and October of 2001 and included interviews with administrators of the New York Networks for School Renewal and its four member organizations; the principals of five schools affiliated with the Networks, two elementary schools and three high schools; a vice president of the New York City Partnership; and two officers of the New York City United Federation of Teachers. Those participating in these interviews were guaranteed anonymity and are thus not identified.
31. *New Visions for Public Schools* Web site, retrieved May 1, 2001 from http://www.newvisions.org/welfram3.html.
32. Center for Educational Innovation-Public Education Association, *Developing Quality Public Education*, n.d.; *CEI-PEA* Web site, retrieved May 18, 2001 from http:www.peaonline.org/home.nsf/vwabout/$first/?open.document.
33. Pearl Rock Kane, "The Difference between Charter Schools and Charterlike Schools," in Ravitch and Viteritti, 65–68; Leanna Stiefel, Robert Berne, Patrice Iatarola, and Norm Fruchter, "High School Size: Effects on budgets and Performance in New York City," *Educational Evaluation and Policy Analysis* 22 (Spring, 2000), 28–29.
34. Institute for Education and Social Policy, *Final Report of the Evaluation of New York Networks for School Renewal, 1996–2001* (New York: Institute for Education and Social Policy, New York University, 2001), 6, 28–32; NYNSR Research Collaborative, *Progress Report on the Evaluation of the New York Networks for School Renewal from July 2000 through January 2001* (New York: Institute for Education and Social Policy, New York University, 2001), 8–12.
35. *Final Report*, 33.
36. New York Networks for School Renewal, *Directory of Participating Schools*, 1999.
37. "What Is Bread & Roses Integrated Arts High School?, n.d. Mimeographed.
38. "The Frederick Douglass Academy Student Creed," 2001. Mimeographed.
39. *New York Times*, January 2, 2001; "The Frederick Douglass Academy," 2001. Mimeographed.
40. *Final Report*, 14, 18, 19, 22, 30, 33.
41. Ibid., 29.
42. Kane, 74–75.
43. *Schools of the 21st Century* Web site, retrieved January 18, 2002 from www.s21c-detroit.org/au_mission.htm.
44. The research reported in this section of the chapter took place in May of 2001 and March of 2002 and involved interviews with administrators of the Schools of the 21st Century Program; the principal, staff facilitator, and two

parent facilitators from one Elementary school in Cluster # 57; and the staff facilitator from another elementary school in the same cluster. Those participating in these interviews were guaranteed anonymity and are thus not identified.

45. Schools of the 21st Century, *New Directions, New Partnerships: A Report to the Community—January 2001*, 11.
46. Schools of the 21st Century, *Able Cluster # 57: Phase IV Implementation Progress Report*, July 1–December 31, 2001, 25–26.
47. *New Directions, New Partnerships*, 2, 6.
48. Schools of the 21st Century, *Leadership Schools: Progress in Implementing Whole School Reform, the Start-Up Year—1999–2000*, December 2000, 3.
49. James P. Comer, "Schools that Develop Children," *The American Prospect On Line* 12 (April 23, 2001) retrieved, December 24, 2004 from www.propsect.org/printfriendly/print/V12/7/comer-j.html; The Yale School Development Program Staff, "Essential Understanding of the Yale School Development Program," in *Transforming School Leadership and Management to Support Student Learning and Development*, ed. Edward T. Joyner, James P. Comer, and Michael Ben-Avie (Thousand Oaks: Crowin Press, 2004), 15–23.
50. Schools of the 21st Century, *Something New Is Happening in our Public Schools: Schools of the 21st Century—The Detroit Annenberg Challenge*, 4.
51. Ibid.
52. *New Directions, New Partnerships*, 5; *Something New is Happening in Our Public Schools*, 7–8.
53. Schools of the 21st Century, *Reflections and Lessons Learned from Schools of the 21st Century*, 25.
54. Detroit Public Schools Performance Report Committee, *1997–98 Detroit Public Schools Performance Report*, 1998, 9, Appendix G-1, G-3; Detroit Public Schools, *2003–04 Annual Report Appendix of Data*, 30–35 retrieved February 20, 2005 from www.detroitk12.org/admin/ceo/pdfs/20032004appendix.pdf.
55. *Reflections and Lessons Learned from Schools of the 21st Century*, 20–21.
56. Ibid., 14.
57. Ibid., 27–32.
58. Ibid., 24, 33.
59. Youth Trust, "Ten Years of Creating Opportunities and Outcomes for Youth, 1989–1999," *Youth Trust* Web site *History*, retrieved March 8, 2001 from www.youthtrust.org/YThist.html.
60. Youth Trust, "*2000*–2001 Minneapolis Public Schools Business and Community Partnerships." Photocopy.
61. The research reported in this section of the chapter involved in-person interviews in March of 2001 with a member of Youth Trust's staff, the liaison between Youth Trust and Pillsbury, a staff member at Jordan Park Middle School involved in its partnership program, four corporate employees who participate in Youth Trust partnerships, and two directors of specific Youth Trust partnership projects. Those participating in these interviews were guaranteed anonymity and are thus not identified.

62. Youth Trust, "Sample Elementary Partnership Models." Photocopy.
63. Youth Trust, "e-Mentoring: Mentoring for the 21st Century," retrieved March 8, 2001 from www.youthtrust.org/hopperarticle.html.
64. Youth Trust, "Workplace Tutoring," retrieved March 8, 2001 from www.youthtrust.org/workplacetutd.html.
65. Youth Trust, "High School Partners," retrieved March 8, 2001 from www.youthtrust.org/highschoolpartners.html.
66. City of Minneapolis, *Census 2000 Information*, retrieved January 30, 2005 from www.ci.minneapolis.mn.us/citywork/planning/census2000.
67. Jordan Park Community School, *Report to the Community*, 2004, retrieved January 25, 2005 from schoolchoice.mpls.mn.us/Jordan_park.html.
68. Larry Cuban, *The Blackboard and the Bottom Line: Why Schools Can't Be Businesses* (Cambridge, MA: Harvard University Press, 2004), 160.
69. Ibid.
70. Ibid., 167–170.
71. I have drawn this account from the following sources: Alex Molnar, *Giving Kids the Business: The Commercialization of America's Schools* (Boulder: Westview Press, 1996), 1–20; Alex Molnar, "Commercial Culture and the Assault on Children's Character," in *The Construction of Children's Character*. Ninety-Sixth Yearbook for the National Study for the Study of Education, Part 2, ed. Alex Molnar (Chicago: University of Chicago Press, 1997), 163–173; Kenneth J. Saltman, *Collateral Damage: Corporatizing Public Schools—A Threat to Democracy* (Latham: Bowman & Littlefield, 2000), xxi–xxv; 57–75; Kenneth J. Saltman, *The Edison Schools: Corporate Schooling and the Assault on Public Education* (New York: Routledge, 2005), 183–184; Melissa K. Lickteig, "Brand-Name Schools: The Deceptive Lure of Corporate-School Partnerships," *The Educational* Forum 68 (Fall, 2003), 44–51; Deron R. Boyles, "The Exploiting Business: School-Business Partnerships, Commercialization, and Students as Critically Transitive Citizens," in *Schools or Markets: Commercialism, Privatization, and School Business Partnerships*, ed. Deron R. Boyles (Mahwah: Lawrence Erlbaum Associates, 2005), 217–240.

6 Educational Partnerships and Community: Education Action Zones and "Third Way" Educational Reform in Britain

1. Daniel T. Rogers, *Atlantic Crossings: Social Politics in a Progressive Age* (Cambridge, MA: Harvard University Press, 1998).
2. Dan Corry, Julian Le Grand, and Rosemary Radcliffe, *Public/Private Partnerships: A Marriage of Convenience or a Permanent Commitment* (London: Institute for Public Policy Research, 1997), 1–3; Bill Coxall, Lynton Robins, and W.N. Coxall, *British Politics since the War* (London: Macmillan Press, 1998).

3. William Richardson, "Employers as an Instrument of School Reform? Education—Business "Compacts" in Britain and America," in *Something Borrowed, Something Learned? The Transatlantic Market in Education and Training Reform* (Washington, DC: Brookings Institution, 1993), 171–192.
4. John Rentoul, *Tony Blair: Prime Minister* (London: Warner Books, 2001), 35–49; Anthony Seldon, *Blair* (London: Free Press, 2004), 32–33, 40. There is some question as how compatible the notion of community that Blair espouses is with the notion of community developed by Macmurray. Compare Blair's understanding of community as it is discussed in this chapter with John Macmurray, *The Structure of Religious Experience* (New Haven: Yale University Press, 1936). For a discussion of this difference see, Ruth Levitas, *The Inclusive Society? Social Exclusion and New Labour* (London: Macmillan Press, 1998), 105–110 and Samuel Brittan, "Tony Blair's Real Guru," *New Statesman* 126 (February 7, 1997), 18–20.
5. Anthony Giddens, *The Third Way: The Renewal of Social Democracy* (Cambridge: Polity Press, 1998), 137.
6. The concept of globalization is used by a variety of modern thinkers to refer to a diverse array of political, economic, and social changes associated with the worldwide economic restructuring that has been occurring since the 1970s. Used in so many different ways for different purposes, it has the features of a floating signifier that I described in chapter 1. I have drawn my understanding of globalization largely from the work of the "third way" thinkers to promote such reforms as partnerships. See Ibid., 30–33; Anthony Giddens, *Beyond Left and Right: The Future of Radical Politics* (Stanford: Stanford University Press, 1994), 4–7; Anthony Giddens, *The Consequences of Modernity* (Stanford: Stanford University Press, 1990), 63–78; James H. Mittelman, *The Globalization Syndrome: Transformation and Resistance* (Princeton: Princeton University Press, 2000), 5–30; Jan Nederveen Pieterse, Globalization as Hybridization," in *The Globalization Reader*, ed. Frank J. Lechner and John Boli (Oxford: Blackwell, 2000), 99–105; Robert Reich, *The Work of Nations* (New York: Alfred A. Knopf, 1991), 110–118; Nikolas Rose, *Powers of Freedom: Reframing Political Thought* (Cambridge: Cambridge University Press, 1999), 142–146.
7. For a discussion of the controversy surrounding the meaning of globalization as well as its very existence see David Held, Anthony McGrew, David Goldblatt, and Jonathan Perraton, *Global Transformations: Politics, Economics and Culture* (Stanford: Stanford University Press, 1999), 1–12.
8. Barry M. Franklin, Marianne N. Bloch, and Thomas S. Popkewitz, "Educational Partnerships: An Introductory Framework," in *Educational Partnerships and the State: The Paradoxes of Governing Schools, Children, and Families*, ed. Barry M. Franklin, Marianne Bloch, and Thomas S. Popkewitz (New York: Palgrave Macmillan, 2003), 6.
9. Giddens, *The Third Way*, 69–86; Anthony Giddings, *The Third Way and its Critics* (Cambridge: Polity Press, 2000), 50–54.
10. Tony Blair, *New Britain* (Boulder: Westview Press, 1997), 32, 51–53, 159–160; Tony Blair, *The Third Way: New Politics for the New Century* (London: Fabian Society, 1998), 4.

11. Tony Blair, "At Our Best When at Our Boldest" (speech presented at the Labour Party Conference, Blackpool, England, October 1, 2002), Retrieved October 19, 2002 from http://www.labour.org.uk/tbconfspeech/
12. Corry et al., 13–15, 57; Peter Jackson, "Choice, Diversity and Partnerships," in *New Labour's Policies for Schools: Raising the Standard*? ed. Jim Docking (London: David Fulton, 2000), 177–190.
13. Blair, *The Third Way*, 4.
14. Department for Education and Employment, *Excellence in Schools* (London: DfEE, 1997), 39; Department for Education and Employment, "Draft of Annual Report on Education Action Zones," (London: DfEE, 2000, mimeographed), 5–6.
15. *Times Educational Supplement*, November 16, 2001.
16. Department for Education and Employment, *Excellence in Cities* (London: DfEE, 1999), 25.
17. *New Britain*, 299.
18. Ibid., 19, 32, 53, 321; Norman Fairclough, *New Labour, New Language*? (London: Routledge, 2000), 127–129.
19. *Excellence in Schools*, 9–14.
20. Ron Letch, "The Role of Local Education Authorities," in *New Labour's Policies for* Schools, 158–176.
21. Levitas, 113–115; see also, Richard Heffernan, *New Labour and Thatcherism: Political Change in* Britain (London: Macmillan Press, 2000); David Howell, *British Social Justice: A Study in Development and Decay* (New York: St. Martin's Press, 1980); Keith Laybourn, *A Century of Labour: A History of the Labour Party, 1900–2000* (Phoenix Mill: Sutton, 2000).
22. *New Britain*, 299.
23. Blair, *The Third Way*, 5–7.
24. Levitas, 58–61.
25. Eva Gamarnikow and Anthony Green, "The Third Way and Social Capital: Education Action Zones and a New Agenda for Educators, Parents, and Community," *International Studies in Sociology of Education* 9 (1999), 3–32; Gewirtz, Sharon, "Education Action Zones: Emblems of the Third Way? *Social Policy Review* 11(1999), 145–165; David Gilborn, "Racism, Poverty, and Parents: New Labour, Old Problems? *Journal of Educational Policy* 13 (1998), 717–735; Sally Power and Sharon Gewirtz, "Reading Education Action Zones," *Journal of Education Policy* 16 (2001), 39–51; Sally Power and Geoff Whitty, "New Labour's Education Policy: First, Second, or Third Way," *Journal of Education Policy*, 14 (1999), 535–546.
26. House of Commons, "Education Action Zones: Meeting the Challenge—The Lessons Identified from Auditing the First 25 Zones," Report by the Comptroller and Auditor General, HC 130 Session 2000–2001, 26 January, 2001, 5–6; "Draft of Annual Report," 6–7.
27. Department for Education and Employment, *Meet the Challenge: Education Action Zones* (London: DfEE, 1999), 5–6, 9–16.
28. Gilborn, 1998; Power and Whitty, 1999; Richard Hatcher, "Labour, Official School Improvements and Equality," *Journal of Education Policy*, 13 (1998),

485–499; Richard Hatcher, "Getting Down to Business: Schooling in the Globalized Economy," *Education and Social Justice*, 3 (2001), 45–59; Stephen O'Brien, "New Labour, New Approach? Exploring Tensions within Educational Policy and Practice," *Education and Social Justice*, 2 (1999), 18–27; Geoff Whitty, "New Labour, Education and Disadvantage," *Education and Social Justice*, 1 (1998), 2–8.

29. Sally Power, Geoff Whitty, Sharon Gewirtz, and David Halpin, "Paving a 'Third Way'? A Policy Trajectory Analysis of Education Action Zones," (Institute of Education, University of London, 2003), Unpublished Paper.
30. Richard Hatcher and Dominique Leblond, "Educational Action Zones and Zone d'Education Prioritaires," in *Education, Social Justice and Inter-agency Working*, ed. Shelia Riddell and Lyn Tett (New York: Routledge, 2001), 33.
31. Socialist Teachers Alliance, *Trojan Horse—Education Action Zones: The Case Against Privatization of Education*, 2nd ed. (Walthamston: Socialist Teachers Alliance, 1998; National Union of Teachers, *An Evaluation of Teachers in Education Action Zones: Executive Summary* (London: National Union of Teachers, 2000).
32. Marny Dickson, Sharon Gewirtz, David Halpin, Sally Power, and Geoff Whitty, "Education Action Zones: Model Partnerships?" in Franklin, Bloch, and Popkewitz, 121–127; Joe Hallgarten and Rob Watling, "Zones of Contention," in *A Learning Process: Public Private Partnerships in Education*, ed. Rachel Lissauer and Peter Robinson (London: Institute for Public Policy Research, 2000), 26–30.
33. Gamarnikow and Green, 1999; Socialist Teachers Alliance, 1998; National Union of Teachers, 2000; Sharon Gewirtz, "Education Action Zones: Emblems of the Third Way? *Social Policy Review*, 11(1999), 145–165; Sally Power and Sharon Gewirtz, "Reading Education Action Zones," *Journal of Education* Policy, 16 (2001), 16–39–51; Sharon Gewirtz, *The Managerial School: Post-welfarism and Social Justice in Education* (London: Routledge, 2002).
34. Sharon Gewirtz, Marny Dickson, and Sally Power, "Government by Spin: The Case of New Labour and Education Action Zones in England," in *Educational Restructuring: International Perspectives on Traveling Policies*, ed. Sverker Lindblad and Thomas Popkewitz (Greenwich: Information Age, 2004), 97–120.
35. Department for Education and Skills, *Education Action Zones: Annual Report 2001* (London: DfES, 2001); Department for Education and Skills, *Education Action Zones—Achievement through Partnerships: The Experience of Education Action Zones—Three Case Studies* (London: DfES, 2003).
36. This research was conducted in January, February, and November of 2001, March of 2002, and December of 2003. These visits included interviews with members of the government's EAZ team, the directors of the North Upton EAZ and two nearby EAZs, head teachers of five of the sixteen schools, a group of six teachers from North Upton, Camden, and East London, a teacher and officer of the North Upton Teachers Association, a parent and governor of one of the zone schools, one of the zone's business partners, and two community activists working in North Upton. Those participating in these interviews were guaranteed anonymity and are thus not identified in this chapter.

37. North Upton is a pseudonym for the real community in which this research was conducted.
38. North Upton (pseudo.), "Education Action Zone Outline Application," April 13, 1999, 6–7; Office of Her Majesty's Chief Inspector of Schools, *Inspection of North Upton (pseudo.) Local Education Authority* (London: Office of Standards in Education, 2000), 1–7.
39. *The Guardian*, March 19, 1999, November 8, 2000, November 11, 2000, November 12, 2000.
40. "Education Action Zone Outline Application," 5.
41. Ibid; North Upton (pseudo.), "EAZ Action Plan," August 8, 2000, 7–9, 41–46.
42. *Times Educational Supplement*, July 24, 1998; *The Guardian*, March 19, 1999, March 20, 1999.
43. *Inspection of North Upton*, 9, 11, 14, 18; *Times Educational Supplement*, November 17, 2000; *The Guardian*, January 18, 1999, March 12, 1999, March 20, 1999, November 17, 2000, November 18, 2000.
44. National Union of Teachers; Socialist Teachers Alliance.
45. *Meet the Challenge*, 3–4; *The Guardian*, January 12, 2000; *Telegraph*, September 6, 1999.
46. Department for Education and Employment, *Education Action Zones (EAZs): Second Application Round Draft Guidance Document* (London: DfEE, 1998), 8, 25–28
47. South East Sheffield, "EAZ Application," January 2000.
48. Hastings and St. Leonards, "Education Action Zone Proposal," 1999, 13.
49. "Education Action Zone Outline Application," 3.
50. Ibid.
51. Ibid, 16–18; "EAZ Action Plan," 60–62.
52. Hallgarten and Watling, 35–36.
53. "Education Action Zone Outline Application," 17.
54. Hallgarten and Watling, 36.
55. This school is a voluntary aided school and while operated and largely supported by the Roman Catholic Church in England it receives some financial support from the government and does not charge its students any fees.
56. National Union of Teachers.
57. Fazal Rizvi and Bob Lingard, "Globalization and Education: Complexities and Contingencies," *Educational Theory* 50 (Fall, 2000), 419–426.

7 SMALLER LEARNING COMMUNITIES AND THE REORGANIZATION OF THE COMPREHENSIVE HIGH SCHOOL

1. U.S. Census Bureau, *American Factfinder*, retrieved September 13, 2008 from http://www.factfinder.census.gov.
2. I have drawn much of this account from a number of regional histories and census reports of the actual city depicted in this chapter by Timberton as well

as documents from the Timberton City Schools. Identifying these sources would reveal the actual name of the city and school, thereby compromising the anonymity that I guaranteed the participants in this study. So, these sources remained unnamed. See also, Timberton [pseudo] City Schools, *Improving Student Achievement through Creating Smaller Learning Communities.* U.S. Department of Education Smaller learning communities Implementation Grant (Timberton: Timberton City Schools, 2003), 8–12; Timberton City Schools, *Enhanced Reading Opportunities/Smaller learning communities Grant* (Timberton: Timberton City Schools, 2006), 2.

3. *Improving Student Achievement through Creating Smaller learning communities,* 50–55; *Enhanced Reading Opportunities/Smaller learning communities Grant,* 3–4; State Office of Education, *Comprehensive School Reform Demonstration Grant Application,* 2002, 4–6.
4. *Improving Student Achievement through Creating Smaller learning communities,* 13; State Office of Education, *Comprehensive School Reform Demonstration Program Local Education Agency Grant Application,* 2002.
5. Southern Regional Education Board, *About SREB,* 1999–2008, 1, retrieved October 5, 2008 from http://www.sreb.org/main/SREB/indes.asp; Southern Regional Education Board, *High Schools That Work: An Enhanced Design to Get All Students to Standards,* n.d., 1–3, retrieved October 4, 2008 from http://www.sreb.org/programs/hstw/pubs/o5vo7_enhanceddesign.pdf.
6. *High Schools That Work,* Ibid.
7. Timberton City Schools, *Smaller learning communities Planning Grant Narrative,* n.d., 2–3, 6, unpublished mimeograph.
8. Walter H. Gaumnitz, *Small Schools are Growing Larger: A Statistical Appraisal* (Washington, DC: GPO, 1959), 1
9. John Rufi, *The Small High School* (New York: Bureau of Publications, Teachers College, 1926), 1–2; Dorothy McCuskey and John Klousia, "What's Right with the American High School?" *The High School Journal* 41 (April, 1958), 298. Edward M. Krug, *The Shaping of the American High School, 1920–1941* (Madison: University of Wisconsin Press, 1972), 46.
10. David Iwamoto, *Small High Schools, 1960–61* (Washington, DC: National Education Association, 1963), 6
11. Ibid., 7; Rufi, 2; Jesse M. Hawley and Irving A. Mather, "The Small High School in California," *California Journal of Secondary Education* 12 (February, 1937), 96; James B. Conant, *The American High School Today: A First Report to Interested Citizens* (New York: McGraw-Hill Book Company, 1959), 77.
12. Rufi, 6; E.H. LaFranchi, "Problems of Small High Schools," *California Journal of Secondary Education* 21 (November 1946), 344–347.
13. Rufi, 86; Hawley and Mather, 97–99; Department of Superintendence, *The Development of the High School Curriculum.* The Sixth Yearbook of the Department of Superintendence, Part I (Washington, DC: National Education Association, 1928), 94; Ray G. Redding, "A 1939-Model Program for One Small School," *California Journal of Secondary Education* 14 (January, 1939), 32–35; Vance D. Lewis, "Variety of Curricular in a Small High School," *California Journal of Secondary Education* 14 (February, 1939), 75–79.

14. Conant, 77–78.
15. See Mark Olssen, John Codd, and Anne-Marie O'Neill, *Education Policy: Globalization, Citizenship and Democracy* (London: Sage Publications, 2004), 22–23.
16. Paul M. Smith, Jr., "The Large or Small High School?," *Journal of Secondary Education* 36 (November, 1961), 389–392.
17. Roger G. Barker and Paul V. Gump, *Big School, Small School: High School Size and Student Behavior* (Stanford: Stanford University Press, 1964), 18–19, 45.
18. Ibid., 41–44, 62.
19. Ibid., 202.
20. Thomas G. Leigh, "Big Opportunities in Small Schools through Flexible-Modular Scheduling," *Journal of Secondary Education* 42 (April, 1967), 175–187.
21. Gaumnitz, Ibid.
22. Iwamoto, 25.
23. For the emergence of this debate during the first half of the twentieth century see Herbert M. Kliebard, *The Struggle for the American Curriculum, 1893–1958*, 3rd. ed. (New York: Routledge, 2004), 1–25. For the more recent history of this debate since 1950 see Barry M. Franklin and Carla C. Johnson, "What the Schools Teach: A Social History of the School Curriculum since 1950," in *The Sage Handbook of Curriculum and Instruction*, ed. F. Michael Connelly (Los Angeles: Sage Publications, 2008), 460–477.
24. Susan F. Semel and Alan R. Sadovnik, "The Contemporary Small School Movement: Lessons from the History of Progressive Education," *Teachers College Record* 110 (September, 2008), 1744–1771; William H. Schubert, "John Dewey as a Philosophical Basis for Small Schools," in *A Simple Justice: The Challenge of Small Schools*, ed. William Ayers, Michael Klonsky and Gabrielle Lyon (New York: Teachers College Press, 2000) 53–66; Rick Ayers, "Social Justice and Small Schools: Why We Bother, Why It Matters." in *A Simple Justice: The Challenge of Small Schools*, ed. William Ayers, Michael Klonsky and Gabrielle Lyon (New York: Teachers College Press, 2000), 95–107; Deborah Meier, *The Power of Their Ideas: Lessons for America from a Small School in Harlem* (Boston: Beacon Press, 1995), 15–38, 47–63; Deborah Meier, *In Schools We Trust: Creating Communities of Learning in an Era of Testing and Standardization* (Boston: Beacon Press, 2002), 25–40, 155–182.
25. Conant, 50–51, 102.
26. For an example of this debate see William R. Johnson, "'Chanting Choristers': Simultaneous Recitation in Baltimore's Nineteenth Century Primary Schools," *History of Education Quarterly* 34 (Spring, 1994), 1–23; William J. Reese, "When Wisdom Was Better than Rubies: The Public Schools of Washington, D.C. in the Nineteenth Century," in *Clio at the Table: Using History to Inform and Improve Education Policy*, ed. Kenneth K. Wong and Robert Rothman (New York: Peter Lang, 2009), 163–164. For a historical account of the ideas of Joseph Lancaster and the impact of his work on schools, see Carl F. Kaestle (Ed.), *Joseph Lancaster and the Monitorial School Movement: A Documentary History* (New York: Teachers College Press, 1973).

27. Joseph P. Nourse, *Class Size and Efficiency in Senior High Schools*, a report to the Superintendent of Schools, May 16, 1928, 1, 15–16, San Francisco Unified School District Records, San Francisco History Center, San Francisco Public Library (hereafter SFSD).
28. Milton B. Jensen and Dortha W. Jensen, "The Influence Class Size Upon Pupil Accomplishment in High-School Algebra I: Summary of Literature and Previous Investigations," *Journal of Educational Research* 21 (February, 1930), 120–137.
29. Milton R. Jensen, "The Influence of Class Size Upon Pupil Accomplishment in High-School Algebra II: The Results of an Investigation in Second-Semester High-School Algebra," *Journal of Educational Research* 21 (May, 1930), 337–356.
30. William S. Vincent, "Dimensions of the Class Size Question," in *Encyclopedia of Educational Research*, 4th ed., ed. Robert L. Ebel (New York: Macmillan, 1968), 143.
31. Barry M. Franklin, "A Historical Perspective on Teaching Low Achieving Children: A First Account," in *Curriculum and Consequence: Herbert M. Kliebard and the Promise of Schooling*, ed. Barry M. Franklin (New York: Teachers College Press, 2000), 135–139.
32. American Federation of Teachers, "Recent Research Shows Major Benefits of Small Class Size," *Education Issues Policy Brief* Number 3 (June, 1998); Elizabeth Graue, Kelly Hatch, Kalpana Rao, and Denise Oen, "The Wisdom of Class Size Reduction," retrieved October 18, 2008 from http://varc.wceruw.org/Sage/AERJGraueetalJuly32006.
33. Tennessee State Department of Education, *The State of Tennessee's Student/Teacher (STAR) Project: Technical Report, 1985–1990*, 181–196. retrieved October 17, 2008 from www.heros-inc.org/star.htm; Health and Education Research Operative Services (HEROS), *Project* Star, 2–3, retrieved October 17, 2008 from www.herosinc.org/star.htm.
34. Mary Anne Raywid, *Taking Stock: The Movement to Create Mini Schools, Schools-Within-Schools, and Separate Small Schools* (New York: ERIC Clearinghouse on Urban Education, Teachers College, Columbia University, 1996), 4–5; Ernest L. Boyer, *High School: A Report on Secondary Education in America* (New York: Harper & Row, 1983), 233–236; John I. Goodlad, *A Place Called School* (New York: McGraw-Hill Book Company, 1984), 309–310, 330–31.
35. San Francisco Board of Education, *The Future of Secondary Education in San Francisco: Some Elements of a Development Plan*, a resource paper for the San Francisco Board of Education and the Advisory Committee for the Master Plan for the Use of Educational Facilities, May, 18, 1972, 1–4, figures 1–2, SFSD.
36. Raywid, 9–13; Patrice Iatarola, Amy Ellen Schwartz, Leanna Stiefel, and Colin C. Chellman, "Small Schools, Large Districts: Small School Reform and New York City's Students," *Teachers College Record* 110 (September, 2008), 1840–1841.
37. *The Power of their* Ideas, 15–63.

38. New York Networks for School Renewal, *Directory of Participating Schools*, September, 1999.
39. Berkeley High School, *Choices, 2007–2008*.
40. Douglas D. Ready and Valerie E. Lee, "Choice, Equity, and the Schools-Within-Schools Reform," *Teachers College Record* 110 (September, 2008), 1931–1932; Linda Shear et al., "Contrasting Paths to Small-School Reform: Results of a 5-Year Evaluation of the Bill and Melinda Gates Foundation's National High School Initiative," *Teachers College Record* 110 (September, 2004), 1931–1932.
41. For a discussion of the various models for downsizing comprehensive high schools and differences and similarities, see Raywid, 21–25 and Valerie E. Lee and Douglas D. Ready, *Schools Within Schools: Possibilities and Pitfalls of High School Reform* (New York: Teachers College Press, 2007), 6–24.
42. *Improving Student Achievement through Creating Smaller learning communities*, 93.
43. [Timbeton Central High School], *Federal Adequate Yearly Progress (AYP) Summary Report*, 2006–2006, 2006–2007, 2007–2008, retrieved March 27, 2009 from state department of Education Web site.
44. Shear et al., 2022; Lee and Ready, 20–21; Patricia A. Wasley et al., *Small Schools: Great Strides* (New York: Bank Street College of Education, 2000), 2; Kathleen Cotton, *New Small Learning Communities: Findings from Recent Literature* (Portland: North-West Regional Educational Laboratory, 2001), 5–6; Linda Christensen, "Rhetoric or Reality? Do Small Schools Change Teaching Practices," *Rethinking Schools* 19 (Summer, 2005), 43–46.
45. "Passion Project," Timberton Central High School, 2007, Mimeographed.
46. Applied Science and Technology Community, Meeting *Agenda*, September 13 2006.

EPILOGUE COMMUNITY IN A COSMOPOLITAN WORLD

1. Malcolm Gladwell, *The Outliers: The Story of Success* (New York: Little, Brown, 2008), 3–7.
2. Ibid, 9.
3. Ibid., 10–11.
4. I have described this shift in the role of the state in examining the mayoral takeover the Detroit Board of Education that constitutes the subject of chapter 4. See Barry M. Franklin, "State Theory and Urban School Reform I: A Reconsideration from Detroit," in *Education under the Security State: Defending Public Schools*, ed. David A. Gabbard and E. Wayne Ross (New York: Teachers College Press, 2004), 117–129.
5. Nikolas Rose, *Powers of Freedom: Reframing Political Thought* (Cambridge: Cambridge University Press, 1999), 174–175.

6. For a good discussion of these contradictory trends and what they mean for the role of the state, see the following articles in a recent issue of *The Economist* 391 (May 9–15, 2009): "A New Pecking Order," 13; "Vive la difference!" (27–29); "There Is No Alternative," (60).
7. For a discussion of the challenges facing the high school see the essays contained in Barry M. Franklin and Gary McCulloch (Eds.), *The Death of the Comprehensive High School: Historical, Contemporary, and Comparative Perspectives* (New York: Palgrave Macmillan, 2007).
8. Robert Fine, *Cosmopolitanism* (New York: Routledge, 2007), ix–x.
9. Ibid., 2.
10. Martha Nussbaum, "Patriotism and Cosmopolitanism," in *For Love of Country: Debating the Limits of Patriotism*, ed. Joshua Cohen (Boston: Beacon Press, 1996), 7.
11. Thomas S. Popkewitz, *Cosmopolitanism and the Age of School Reform: Science, Education, and Making Society by Making the Child* (New York: Routledge, 2008), xiii, 4, 111–112.

Index

100 Black Men, 92

academic achievement, 32–3, 43, 57–8, 69, 70, 73, 76, 79, 81, 85, 107, 117–18,136, 149, 151, 153–4, 158, 168, 170, 176–7, 178, 193, 194, 200–4
Academy of the Americas (Detroit), 128–9, 131
Achieve!Mpls (Mineapolis), 135, *See under* Youth Trust
Action Forum, 150, 151, 156, 158, 159, 161, 162–3, 164
Adamany, David, 95
Ad Hoc Committee of Citizens Concerned with Equal Educational Opportunity, 70
African American Men's Organization, 93
Afro-American Teachers Association, 49
Agbemakplido, Wilhemina, 18
All Day Neighborhood Schools (ADNS) (New York City), 37
Alsop, Joseph, 43
Alvarado, Anthony, 13
American Federation of Teachers, 44
Anderson, Benedict, 6
Anderson, Yvette, 91
Annenberg Challenge, 105, 113–16, 119–20, 131–4, 140, 170, 187
Annenberg, Walter, 113
Archer, Dennis, 82, 87–9, 90, 93
Ardrey, William, 62
Arwood, E. Dennis, 65
Association of Community Organizations for Reform Now (ACORN), 105, 108, 109, 110, 111, 112, 121, 140
Augustine, Saint, 7

balanced literacy, 13–14
Baltimore, Maryland, 33
Barker, Roger, 182–3
Basile, Patrick, 72
Batchelor, Michael, 75
B Curriculum Experiment, 185–6
Beagle, Si, 37, 51
Beard Elementary School (Detroit), 128–9, 131
Beckham, Bill, 93
Bedford-Stuyvesant (New York City), 45–6, 50–1
Bellah, Robert, 18
Benjamin Franklin High School (New York City), 21–2
Bennett, Loren, 86
Benson, Lee, 18, 21
Beresin, Alan, 13
Berkeley High School (Berkeley, California), 187–8
Booker, Cory, 7
black identity, 73–8
black nationalism, 24, 55, 71
black power, 24, 55
Blair, Tony, 144, 145, 146, 147, 155
Blouse, Richard, 86
Blunkett, David, 154
Bradfield, Horace, 66
Bread and Rose Integrated Arts High School (New York City), 116–17, 122
Brown, Aaron, 38

Brown, H. Rap, 74
Brown, Nathan, 38
Browne, R. J., Jr., 65
Brownell, Samuel, 57, 59, 61
Brown v. Board of Education, 54
Burgos, Rosa, 9
Burnley, Kenneth, 95–8, 99
Byers, Stephen, 146

Carlen, B. A., 66
Carmichael, Stokely, 74
Carson, Robert, 53–4
Carty, Arthur, 57, 59, 61, 62
Center for Collaborative Education (CCE) (New York City), 115, 121
Center for Community Partnerships (Philadelphia), 23
Center for Educational Innovation (CEI) (New York City), 115, 121
Center for Urban Education (New York City), 36, 39–44, 48
Central High School (Detroit), 62
centralization, 30, 78, 91
Central Park East Secondary School (New York City), 115, 187
Chambliss, Earl, 82
charter schools, 105, 106, 107–13, 134, 199
Citizens Advisory Committee on Equal Educational Opportunity, 64–7
Citizens Advisory Committee on School Needs, 63–5
civic capacity, 11, 12, 99, 113, 121, 122, 125, 131, 133, 168–9, 172, 208
civil society, 3, 103, 106, 145
Clapp, Elsie, 21, 23
Clarke, Lewis, 51
Clark, Kenneth, 66–7
Cleage, Albert, 72–4
Clinic for Learning (New York City), 25, 32, 45–55, 102, 103, 106
  conflict over, 47–52
  demise, 51–2
  established, 45–7
  evaluation of, 48–9, 50–1
Cogan, Charles, 39
Cohen, David, 29–31, 45
Colding, Charles, 58, 60, 62
collective belonging, 9, 12, 14, 23, 55, 99, 140, 172, 205–6, 208, 210, 212
Comer School Development Program, 127
common good, 7, 31, 101, 140, 145, 167, 205, 210, 211
community, xiii, 2–3, 4–5, 6, 7, 8–9, 10, 12, 15, 18, 25, 31, 55, 58, 78, 82, 99, 101, 103, 112, 113, 121, 124, 131, 139, 144, 145, 167, 168–9, 172–3, 175, 195–7, 205, 208, 210, 211, 213
Community Advisory Council (New York City), 21–2
community control, 30, 32, 71–4, 77
compensatory education, 29, 33, 43, 70
Conant, James Bryant, 64, 181–2, 184
Congress of Racial Equality (CORE), 40, 53–4
Conservative Party (UK), 144, 147, 151, 165
Cooley, Charles Horton, 10
cosmopolitanism, 212
Covello, Leonard, 21–2, 23
Cuban, Larry, 139
Cullen, Claudia, 61
curriculum, xiii, 3, 13, 53, 59, 61, 63, 75, 181, 184
  Annenberg Challenge (Detroit), 128, 132, 134
  Annenberg Challenge (New York City), 116–18, 118–20
  Clinic for Learning (New York City), 46–7, 51–2
  curriculum differentiation, 63–7, 69, 70, 76
  Education Action Zones (UK), 150, 155, 157–8, 169–70, 178
  More Effective Schools (New York City), 35–6, 43–4
  smaller learning communities, 189–90, 193–204

Daley, Richard, 81
Davis, William, 7
De Alva, Jorge Klor, 78
Dean, Mitchell, 26

decentralization, 30, 32, 79, 91
DeGrow, Dan, 83, 84, 87, 88
Department for Education and Employment (DfEE) (UK), 161
Department for Education and Skills (DfES) (UK), 152
Detroit Association of Black Organizations, 92
Detroit Board of Education, 57, 59, 67–8, 82, 84, 84, 87, 102, 122
Detroit Education Association (DEA), 65
Detroit Federation of Teacher (DFT), 65, 92, 95
Detroit Free Press, 61, 65, 75, 83, 89, 90, 91
Detroit High School Study Commission, 70
Detroit (Michigan), 25, 32, 44, 55, 57–79, 80–103, 105, 106, 107, 125–34
Detroit News, 84, 86, 88, 91, 96, 97
Detroit Organization of School Administrators and Supervisors, 92
Detroit Recorders Court, 83
Detroit Regional Chamber of Commerce, 92
Dewey, John, 10, 15–21, 24, 115, 184
Dinkins, David, 111
discourse, xiii, 8, 99, 101–2, 182–4, 186, 205, 210, 213
Donaldson, George, 59
Donovan, Bernard, 31, 37, 42, 47, 50–1, 53, 54

Eastern High School (Detroit), 60
Ecorse (Michigan), 44
Ecumenical Ministers Alliance (Detroit), 92
Edison Schools, 108, 109, 110, 140, 155
Education Action Zones (EAZ) (UK), 25, 146, 149–73
Ehrenhalt, Alan, 4
Eiges, Hubert, 66
Elementary and Secondary Education Act, 8
Ellis, Art, 89
Ellison, Ralph, 54
empty signifier, 9, 98, 205, 212
Engler, John, 81, 87, 89, 90, 93, 94
Excellence in Cities Program (UK), 147, 152

Fege, Arnold, 8
Fine, Sydney, 58
Finney, Ross L., 10
Fisher, Frank, 26
Flint (Michigan), 44
floating signifier 9
Ford Foundation, 32, 33, 45, 106
Foucault, Michel, 26
Fowell Junior High School (Minneapolis), 185–6
Fowler, Robert Booth, 4, 58, 98
Fox, David, 40, 41
Fraser, Elizabeth, 78
Frederick Douglas Academy (New York City), 117
Fuller, Bruce, 15

Garrett, Nathan, 66
Garrison, Lloyd, 37, 39
Garvey, Marcus, 71
genealogy, xiii, 26
Giardino, Alfred, 38
Giddens, Anthony, 144
Giuliani, Rudolph, 105, 108, 109
Gladwell, Malcolm, 209–10
Glaude, Eddie, 24
globalization, 140, 144, 145, 172, 208, 210, 211, 212
Gracie, David, 57
Grandholm, Jennifer, 97
Gray Areas Program, 32, 106
Great Cities School Program, 32
"Great Community," 16–17, 24
Greeley, Kathy, 8
Green, Eddie, 85, 90
Green, Malice, 83
Gregory, Karl, 57, 71
Gross, Calvin, 35
Gump, Paul, 182–3

Hallgarten, Joe, 162
Hall, Vangilee, 39
Harkavy, Ira, 18, 22
Harper, William Rainey, 21
Hegel, Newton, 185
Herbert, Solomon, 54
Higgenbotham Elementary School (Detroit), 60, 69
High Schools that Work, 177–8
Hillery, George, 5
history of the present, xiii, 26
Hollander, Florence, 66
Hubbard, Lea, 13, 14

Inner City Parents Council (Detroit), 74–5
Institute of Learning (New York City), 53–4
intermediary organization, 114–16, 121
Intermediate School (IS) 201 (New York City), 52–3
International Monetary Fund, 144
Irwin, Frances, 86

Jackson, Robert, 65
Jackson, Thomas 86
Jensen, Dortha, 185
Jensen, Milton, 185
Jones, Sterling, 86
Jordan Park Community School (Minneapolis), 136–9
Julian Union High School District (California), 183
Julia Richmond Education Complex (New York City), 179

Kaufman, Nathan, 67
Keller, Suzanne, 5
Kilpatrick, Bernard, 93
Kilpatrick, Carolyn, 93
Kilpatrick, Kwame, 89, 93, 97, 98
King, Martin Luther, Jr., 7, 74
Kluger, Richard, 54
Knudsen Junior High School (Detroit), 73

Labaree, David, 99–102
Labour Party (UK), 25, 143, 144, 145, 146–50, 151,152, 154, 160, 162, 165, 167, 168, 169, 170, 171
LaFranchi, E.H., 181
Lake, Henry, 66
Leland, Burton, 83, 86
Lemmons, Lamar, 89
Levy, Howard 105, 108, 109, 111
Lewis, Charles, 62
Lindsey, John 38, 42
Lippmann, Walter, 17
local education authority (UK), 146, 154, 161, 162, 165
Logan Elementary School (Detroit), 128–9, 131, 132

MacIver, Robert, 2–3
Mackenzie High School (Detroit), 66
Maher, Bill, 7
Malcolm X, 115
Marburger, Carl, 72
Martinez, Juan Jose, 83
mayoral takeover of the Detroit Board of Education, 25, 79, 81–97, 99–102, 133
McGlobin, Sophie, 72
McMurray, John, 144
Mead, George Herbert, 10
Mehan, Hugh, 13
Meier, Deborah, 114, 121, 187
Michigan Chronicle, 71, 73, 91
Milwaukee (Wisconsin), 33, 44
Minneapolis (Minnesota), 25, 44, 105, 106, 114, 134–40, 170, 185
Mirel, Jeffrey, 58
Moore, Claudia, 67
More Effective Schools (New York City), 25, 29–45, 54–5, 102, 103
  conflict over, 37–45
  demise, 44–5
  establishment, 32–6,
  evaluation, 36–7, 40–1, 44
Muhammad, Elijah, 71
Mumford High School (Detroit), 96

National Association for the Advancement of Colored People (NAACP), 57, 69, 70, 92
National Curriculum (UK), 150, 151, 155, 157, 160, 165, 166
National Education Association, Commission on Professional Rights and Responsibilities, 57–8
National Teachers' Pay and Conditions Document (UK), 150, 160, 166
National Union of Teacher (NUT) (UK), 166
New Detroit, 85, 93
New Orleans Recovery School District, 14–15
New Visions for Public Schools (New York City), 115, 121, 123
New York City Board of Education, 33, 45, 110, 111, 122, 123
New York City (New York), 13, 21, 22, 25, 29–55, 102, 103, 105, 106–25, 133, 134, 139, 140, 167, 170, 171, 179, 187
New York Networks for School Renewal (NYNSR) (New York City), 106, 109, 114, 121, 122, 125,187
New York University, 45, 48, 106
No Child Left Behind, 8
Northeastern High School (Detroit), 69
Northern High School (Detroit), 25, 57–63, 68, 75–7, 78
North Upton, Borough of (pseudo.), 153–67
Northwestern High School (Detroit), 68

Office for Standards in Education (OFSTED) (UK), 148, 154, 166

Partnerships, 11, 30, 34, 103, 105, 106, 108, 112, 113–16, 119, 122, 129, 132, 135, 136, 140, 143, 144, 145–6, 148, 149, 152, 155, 160, 165, 170–2
Pattengil Elementary School (Detroit), 60
Person, Leontine, 88
Peters, Gary, 88, 89
Pettigrew, Thomas, 43
Phillips, Derek, 6
Pillsbury Company, 135
Plant, Raymond, 8, 9
Podair, Jerald, 77
policy narrative, xiv, 26–7
Post Junior High School (Detroit), 60, 76
privatization, 108–9, 110, 140, 151, 155, 165, 171
Proposition A (Michigan), 82
"public," 16, 18
Public Act 10 (Michigan), 89–90, 94
Puckett, John, 18
Putnam, Robert, 10
Powell, Colin, 24

racial conflict, 24, 25, 55, 206–7
  New York city, 42–3, 52–3
  Detroit, 68–71, 73, 86, 90–4, 99
Randolph, A. Phillip, 35
Redford High School (Detroit), 58
Redmond, Daryl, 83, 84
reflex arc concept, 17–18
Rice, Condeleezza, 24
Richards, Paul, 61
Robertson, John, 47–8, 51
Robinson, Remus, 67–8
Rogers, David, 143
Rosario, José, 12
Rose, Nikolas, 4, 211
Roseto (Pennsylvania), 209–10
Ross, Edward L., 10

St. Joseph's Church (Detroit), 57, 59, 74
Samples, James, 62
San Diego Unified School District (California), 13, 180
San Francisco Board of Education (California), 184–5, 186
school desegregation, 54, 66, 67
School District of Philadelphia (Pennsylvania), 22
school integration, 12, 31, 35, 36, 40, 43, 52, 53, 55, 65, 67, 68, 71, 75, 77, 78

school segregation, 24, 43, 54, 58, 60, 65, 66, 67, 68, 71
Schools of the 21st Century (Detroit), 10, 126–34
School Under Registration Review (SURR) (New York), 107
Schwager, Sidney, 38, 39, 41
Schwartz, Louis, 45, 47–9, 51
Schwartz, Robert, 43
Scott, Thomas, 58
Scruggs, Ramon, 67
Shagaloff, June,, 70–1
Shanker, Al, 37, 39
Shapiro, Rose, 31
Sharpton, Al, 111
Shields, Glen, 64
Shorris, Earl, 78
small classes, 34, 184–6
small schools, 115–17, 124, 179–87, 205
smaller learning communities, 1–2, 25, 175–8, 180, 188–99, 200–8
Smith, Marshall, 43
Smith, Paul, 182
Smith, Virgil, 87, 93
Snedden, David, 10
social capital, 10, 11, 12, 145
social inclusion, 149, 150, 152, 159
Southeastern High School (Detroit), 60
Spaulding, W.B., 66
specialist schools (UK), 148, 150, 151, 155, 160
special service schools (New York City), 33, 46
"stakeholder society," 148–9
Stanford University School Redesign Network, 202
state, 3, 7, 25, 81, 82–3, 84, 85, 86, 91, 94, 102, 103, 106, 107, 123, 140, 141, 144, 145, 146, 147, 149, 151, 165, 170, 177, 208, 210–11, 212
Stein, Mary Kay, 13
Stone, Clarence, 11
Stroud, Joe, 90–1

Tennessee Star Project, 186
"third way," 144, 145, 151, 158, 167, 171, 172, 210
Thorndike, Edward L., 10, 18
Thurgood Marshall High School (New York City), 124–5
Timberton Central High School (pseudo.), 1, 175, 177–208
Timberton, (pseudo.), 1, 175–7
Tyler, Ralph, 22

United Auto Workers, 69
United Federation of Teachers, 29, 32, 33, 34, 35, 41, 52, 110
United Negro Improvement Association, 71
University of Michigan, 17, 57, 68
University of Pennsylvania, 22–3
Urban League, 64–5. 69, 92

Vallas, Paul, 14–15
Vaughn, Ed, 93

Waldmeir, Peter, 88
Walker, Judy, 75
Watling, Rob, 162
Watson, John, 18
Weaver, R.H., 65
Weeks, George, 88
Wells, Charles, 70
West, Cornell, 77–8
West Philadelphia Improvement Corps (WEPIC) (Philadelphia), 22–3
Whitelaw Reid Junior High School (JHS # 57) (New York City), 32, 45–52, 175
Whittle, Christopher, 112–13
Wieder, Patricia, 61
Williams, Angelia, 86
Williams, Raymond, 9
Wills, Millicent, 62
Wolf, Steward, 209–10

Young, Coleman, 82, 87
Young, Leonard, 92
Youth Trust (Minneapolis), 106, 134–40, 140, 170, *see also* under Achieve!Mpls